SUSTAINABLE AUTOMATED AND CONNECTED TRANSPORT

TRANSPORT AND SUSTAINABILITY
VOLUME 19

SUSTAINABLE AUTOMATED AND CONNECTED TRANSPORT

EDITED BY

NIKOLAS THOMOPOULOS
University of Surrey, UK

MARIA ATTARD
University of Malta, Malta

and

YORAM SHIFTAN
Israel Institute of Technology, Israel

The book has been supported by the Action CA16222 of the European Cooperation in Science and Technology (COST) entitled "*Wider Impacts and Scenario Evaluation of Autonomous and Connected Transport*" (WISE-ACT). WISE-ACT offered the opportunity for multiple researchers from different countries with diverse backgrounds and expertise to explore relevant policy implications of autonomous and connected vehicles, and identify required policy developments at global level.

COST (European Cooperation in Science and Technology) is a funding agency for research and innovation networks. Our Actions help connect research initiatives across Europe and enable scientists to grow their ideas by sharing them with their peers. This boosts their research, career and innovation.

www.cost.eu

Funded by the European Union

United Kingdom – North America – Japan
India – Malaysia – China

Emerald Publishing Limited
Emerald Publishing, Floor 5, Northspring, 21-23 Wellington Street, Leeds LS1 4DL.

First edition 2024

British Library Cataloguing in Publication Data
A catalogue record for this book is available from the British Library

ISBN: 978-1-80382-350-8 (Print)
ISBN: 978-1-80382-349-2 (Online)
ISBN: 978-1-80382-351-5 (Epub)

ISSN: 2044-9941 (Series)

Printed and bound by CPI Group (UK) Ltd, Croydon, CR0 4YY

INVESTOR IN PEOPLE

CONTENTS

LIST OF FIGURES AND TABLES

Figures

Tables

ABOUT THE CONTRIBUTORS

Serio Agriesti is a PhD Student at Aalto University. His previous work focused on the impact assessment of innovative transport systems such as Connected and Automated Vehicles (CAVs) and Truck Platooning. His current research activities focus on agent-based modelling, performance evaluation and fleet sizing for CAVs.

Constantinos Antoniou (PhD from MIT, 2004) is a Full Professor and Chairholder at the Chair of Transportation Systems Engineering at the Technical University of Munich (TUM), Germany. His research focuses on modelling and optimisation, data analytics and machine learning, and human factors for future mobility systems. He has authored more than 450 publications, including 185 journal papers.

Maria Attard is Head of Geography and Director of the Institute for Climate Change and Sustainable Development at the University of Malta. She is Co-editor of Research in Transportation Business and Management, Associate Editor of Case Studies on Transport Policy and sits on the editorial board of the *Journal of Transport Geography*. Between 2002 and 2008, she was a consultant to Malta's government and helped develop the first white paper on transport policy (2004) and implement the 2006 Valletta Strategy including park-and-ride, pedestrianisation and road pricing (2007). She also supported the planning for the 2011 public transport reform. She sits on the Steering Committee of the WCTR and is a Cluster co-chair for NECTAR.

Eran Ben-Elia is an Associate Professor at the Department of Environmental, Geoinformatic and Urban Planning Sciences, Ben-Gurion University of the Negev, Head of the GAMES Lab and the CITI Center for Intelligent Transportation & Innovation. His research interests lie in applications of emerging technologies, including automated vehicles, to promote sustainable mobility futures using game-based methods, agent-based simulation models, big data analytics and AI.

Eda Beyazit is a Research Fellow in Active Travel and Micro-mobility at the Centre for Transport & Society, University of the West of England, Bristol, UK. She previously worked as an Associate Professor in Urban Planning at the Faculty of Architecture, Istanbul Technical University, and founded the IstanbulON Urban Mobility Lab. Eda was an Urban Studies Foundation International Fellow at the University of Manchester between 2022 and 2023. Eda completed her DPhil in Transport Geography at the Transport Studies Unit, University of Oxford, focusing on transport-related socio-spatial inequalities. Lately, her interests include gender, informality, and the periphery. Eda has chaired the Working Group 2 of WISE-ACT alongside Karel Martens.

Pablo Cabanelas is an Associate Professor in the Department of Business Organization and Marketing at University of Vigo. He is Principal Investigator in the Knowledge Organization Research Group and Leader in several projects. His research interests are networks, industrial marketing, competitiveness, business models and mobility. His research has been published in more than 30 publications in journals and books.

Biagio Ciuffo received a PhD. degree in Transportation Engineering from the Department of Transportation Engineering, University of Napoli Federico II, in 2008. He held a three-year post-doctoral position at the European Commission Joint Research Centre (JRC), Ispra, Italy, working on the sustainability assessment of traffic and transport-related measures and policies. He is currently an Official of the European Commission, working for the Directorate for Energy, Transport, and Climate of the JRC. In the past, he has led different projects concerning the analysis of the environmental and economic impacts of different transport policies. He is also leading the JRC Project focusing on the wide implications of connected and automated mobility. He has published more than 100 scientific papers in peer-reviewed journals and conference proceedings in transportation and traffic engineering, and he has been awarded several international prizes for his research. He is the Deputy Editor-in-Chief of the *IET Intelligent Transport Systems Journal* and an Associate Editor of the *IEEE Transactions on Intelligent Transport Systems*.

Mor Collins was awarded an MA in Geoinformatics at the Department of Environmental, Geoinformatic and Urban Planning Sciences, Ben-Gurion University of the Negev. He specialises in advanced geographic information systems analytics, simulation and game-based approaches. He currently is working as a Project Development Manager at Enlight Renewable Energy Ltd (ENLT).

Henriette Cornet is Senior Manager for Shared Automated Mobility at the International Association of Public Transport (UITP). She received her Doctoral degree from TU Munich in 2012, and worked for the automotive industry as a Consultant in Germany and as Principal Investigator in the research institute TUMCREATE in Singapore.

Nima Dadashzadeh is a Lecturer (Assistant Professor) in Transport and Business Analytics at the Huddersfield Business School, University of Huddersfield, UK. He is an Executive Committee Member of the Universities' Transport Study Group (UTSG) which includes 60 universities across the UK. He was also a Management Committee (MC) Member from Slovenia in EU COST Action CA16222 (WISE-ACT): Wider Impacts and Scenario Evaluation of Autonomous and Connected Transport. His research focuses on traffic simulation, transport planning and modelling, accessibility, and social inclusion of emerging transportation systems such as Autonomous and Connected Transport (ACT), Mobility-as-a-Service (MaaS), and Shared Mobility.

Arnór B. Elvarsson is a Research Associate at ETH Zürich. His research focuses on infrastructure planning under uncertainty for transportation systems, including scenario generation, process development and infrastructure system modelling to support decision-making for future-proof infrastructure.

Shelly Etzioni was awarded her PhD degree in Civil and Environmental Engineering from Technion, Israel Institute of Technology, focusing on automated and shared automated vehicles and their acceptance, using advanced modelling techniques. She is currently working as a modelling consultant for the Israeli Ministry of Transport.

Alexander Fürdös holds a Master's degree in Urban and Regional Planning and has studied law in Vienna and Linz. His professional experience of more than 25 years covers project management of national and international transport and traffic planning projects and studies. At present, he is a Senior Expert at AustriaTech and a specialist in traffic safety; platooning and legal questions related to automated mobility.

Nikolaos Gavanas is Assistant Professor and Director of the Research Unit of Infrastructure, Technology Policy and Development, Department of Planning and Regional Development, University of Thessaly, Greece. He has participated as a member in the Sectoral Scientific Committee of Environment, Energy and Sustainable Mobility of the National Council for Research, Technology and Innovation. He has worked as a Policy Officer in DG RTD.

Wolfram Klar holds a Master's degree in Spatial and Regional Planning and has experience in being a consultant and transport engineer with involvement in various projects in the field of transport and telematics. He previously worked as a Project Manager for road telematics and was responsible for the implementation of nationwide tolling systems. At present, he is a Senior Expert at AustriaTech and Head of the team Automated Mobility & Safety.

Anastasios Kouvelas is the Director of the Traffic Engineering and Control Research Group at the Institute for Transport Planning and Systems (IVT), Dept. of Civil, Environmental and Geomatic Engineering, ETH Zurich (since August 2018). Prior to joining IVT, he was a Research Scientist at the LUTS, EPFL (2014–2018), and a Postdoctoral Fellow with PATH at the University of California, Berkeley (2012–2014). He is a Senior Member of IEEE.

Michail A. Makridis is a Senior Research Scientist and Deputy Director of the Traffic Engineering Group (SVT) of ETH Zürich, Switzerland. Before, he was the scientific responsible for the Traffic Modelling Group of the Sustainable Transport Unit at the Joint Research Centre (JRC) of the European Commission (EC). He holds a PhD. in Computer Vision from the Democritus University of Thrace, Greece. His interests are in Traffic Flow, Management, and Control for future Intelligent Transportation Systems in the presence of Connected and

Automated Vehicles. In 2022, he received the JRC Annual Awards on Excellence in Research from the European Commission. He is a Member of IEEE.

Karel Martens is a Professor of Transport Planning at the Faculty of Architecture and Town Planning, Technion – Israel Institute of Technology where he holds the David J. Azrieli Chair in Architecture and Town Planning. He is an expert on the nexus between transport, justice and equity and has published extensively on the topic. Together with Eda Beyazit, he co-chaired Working Group 2 of WISE-ACT, which explored the potential social challenges of autonomous and connected transport.

Konstantinos Mattas received a Civil Engineering degree, specialised in transportation, the M.Sc. degree in applied mathematics from the Democritus University of Thrace, Xanthi, Greece, and a PhD degree from the Democritus University of Thrace, working on fuzzy logic applications in transportation engineering. Since 2017, he has been working at the European Commission Joint Research Centre. His research interests include intelligent transportation systems, simulation of vehicle dynamics and driver behaviour, microscopic simulation of traffic networks and network control, optimisation, and traffic safety.

Alexandros Nikitas is a Professor in Sustainable Transport Futures for Huddersfield Business School and Huddersfield's Future Mobility Lab Founding Director. He is the Chair of the Universities' Transport Study Group's Executive Committee; a group that is the premier academic forum in the UK and Ireland for transport research with 60 Universities on board. Prof. Nikitas is working on the agendas of sustainable and socially inclusive mobility and automated, connected, alternatively fuelled and shared transport and has published already more than 130 peer-reviewed research outputs. He is the Associate Editor of the *Journal of Transport & Health* and a member of the Editorial Boards of the *Journals Transportation Research Part A: Policy and Practice, Research in Transportation Business & Management and Case Studies on Transport Policy*. He is a Visiting Professor at Chalmers University of Technology, Sweden and at the National Technical University of Athens, Greece. Prof Nikitas is also a world's top 2% researcher for the years 2020, 2021 and 2022 according to the Stanford/Elsevier ranking list (c-score based).

Daniela Paddeu is an Associate Professor in Sustainable Freight Futures and freight specialist at the University of the West of England, Bristol, UK. She has almost 10 years' of experience in designing and delivering research projects on innovative and sustainable freight solutions in the UK and Europe, including stakeholder engagement, co-design, automation, and decarbonisation. Daniela has significant experience in enabling stakeholders with different backgrounds and experiences to share views and design appropriate solutions.

Graham Parkhurst is a Professor of Sustainable Mobility and Director of the Centre for Transport & Society, University of the West of England Bristol,

UK added. Graham has taught and researched transport policy since 1991. His current research examines the sustainable mobility implications of greater automation and electrification, taking a critical lens to the discourses and practices of "future mobility." Graham is a Fellow of the Royal Geographical Society.

Claudio Roncoli is an Associate Professor of Transportation Engineering at Aalto University, Finland. His research interests include real-time traffic management, modelling, optimisation, and control of traffic systems with connected and automated vehicles, as well as smart mobility and intelligent transportation systems. He is the author of more than 90 peer-reviewed scientific publications.

Hashmatullah Sadid is a Research Associate and a PhD Candidate at the Technical University of Munich (TUM), Germany. His research focuses on ACT including simulation-based evaluation of ACT deployment scenarios and the use of deep learning methods and open-source data for capturing the driving behaviour of autonomous vehicles.

Dominik Schallauer works at AustriaTech as an Expert in the field of Automated Mobility & Safety. He holds a Master's degree in Environment and Bio-Resources Management, where he specialized in transport and mobility.

Yoram Shiftan is a Professor at the Faculty of Civil and Environmental Engineering, Technion, Israel Institute of Technology and Head of the Israel Smart Transportation Research Center (ISTRC). His research interests include understanding sustainable futures of automated mobility using advanced choice modelling and other emerging methods.

Emily Soh is an Urban Researcher who has worked on cross-cutting urban issues including ageing and the city, transport, and food systems. Her research interests revolve around urban systems transitions, with an aspiration of guiding them toward public benefit and greater equity. She completed her PhD at the Technion – Israel Institute of Technology where she researched the public values and the equity dimension of autonomous vehicle implementation. She is currently a Postdoctoral Fellow at the National University of Singapore, working on food systems transitions for small open economies.

Aggelos Soteropoulos is a Senior Expert for Automated Mobility & Safety at AustriaTech. Before joining AustriaTech, he worked as a Researcher at the Institute for Transportation System Planning at the Technical University of Vienna and the Austrian Road Safety Board and finished his doctoral thesis on the effects of automated vehicles on mobility and urban development in 2022.

Nikolas Thomopoulos is an Associate Professor in Transport at the University of Surrey. He led the University of Surrey Living Lab and currently is the Sustainable Transport & Mobility Programme Co-Leader at the Institute for Sustainability.

Previously, he was the Chair of WISE-ACT, an international network of 200 experts in 42 countries, which focused on the wider impacts and scenario evaluation of automated and connected transport. His current research focuses on co-creating innovative and inclusive new mobility services.

Lena Zeisel holds a Master's degree in Eco Design and is working as an expert at AustriaTech, amongst others on the topic of Automated Mobility. Within the field of Automated Mobility, her interest lies especially in sustainability and inclusion aspects.

ACKNOWLEDGEMENTS

The book has been supported by the Action CA16222 of the European Cooperation in Science and Technology (COST) entitled "*Wider Impacts and Scenario Evaluation of Autonomous and Connected Transport*" (WISE-ACT). WISE-ACT offered the opportunity for multiple researchers from different countries with diverse backgrounds and expertise to explore relevant policy implications of autonomous and connected vehicles, and identify required policy developments at global level.

COST (European Cooperation in Science and Technology) is a funding agency for research and innovation networks. Our Actions help connect research initiatives across Europe and enable scientists to grow their ideas by sharing them with their peers. This boosts their research, career and innovation.

www.cost.eu

This book could not exist without our authors who have so willingly contributed their knowledge. We also thank the reviewers who supported the peer-review process undertaken to make this volume a success.

More information about WISE-ACT can be found at https://wise-act.eu/

cost EUROPEAN COOPERATION IN SCIENCE & TECHNOLOGY

Funded by the European Union

PART ONE

ACT INNOVATION

CHAPTER 1

INTRODUCTION TO SUSTAINABLE AUTOMATED AND CONNECTED TRANSPORT

Nikolas Thomopoulos[a], Maria Attard[b], Yoram Shiftan[c] and Lena Zeisel[d]

[a]*Department of Tourism and Transport, University of Surrey, UK*
[b]*Institute for Climate Change and Sustainable Development, University of Malta, Msida, Malta*
[c]*Faculty of Civil and Environmental Engineering, Technion Israel Institute of Technology, Haifa, Israel*
[d]*AustriaTech Federal Agency for Technological Measures, Vienna, Austria*

ABSTRACT

The 26th United Nations Climate Change Conference of the Parties (COP26) has reinvigorated the policy focus on sustainable transport. Automated and Connected Transport (ACT) has been featured as a promising technology-based option to aid in meeting the Sustainable Development Goals (SDGs). Despite progress in certain areas of sustainability, there are still a lot of SDGs where limited progress has been observed since the 2015 Paris Agreement, particularly regarding the social pillar of sustainability which is reflected from the user perspective. This chapter will set the scene for this edited volume first by contrasting ACT potential with the SDGs and then by highlighting the requirement to focus more on addressing user needs through ACT. Remarkably, scholars have been increasingly sceptical about the transition to fully automated and connected vehicles, thus it is pertinent to highlight relevant opportunities and risks. Chapter recommendations foster the promotion of a Quadruple Helix approach to operationalise the inclusion

Sustainable Automated and Connected Transport
Transport and Sustainability, Volume 19, 3–22

Published under exclusive licence by Emerald Publishing Limited
ISSN: 2044-9941/doi:10.1108/S2044-994120240000019001

of social concerns (e.g. gender balance and equity) in Sustainable Urban Mobility Plans (SUMP) across the world.

Keywords: Automated and connected transport; autonomous; living lab; quintuple helix; responsible innovation; sustainable development goals

INTRODUCTION

Following the COVID-19 shock to transport systems worldwide due to restrictions and lockdowns, the 26th United Nations Climate Change Conference of Parties (COP26) held in Glasgow in late 2021 reinvigorated the policy focus on sustainable transport. The summit was supposed to be the 5-year checkpoint after the historical 2015 climate agreement signed in Paris at COP21. Indeed, COP21 has been celebrated as the first step of a major success for sustainable development, since scientists and researchers have been raising the alarm with policymakers, industry and citizens for a number of years about the irreversible impact of human activities on climate. As a result, the 17 SDGs were also agreed at the 2015 COP21 and projected through the 2030 Agenda for Sustainable Development (UN-IGS, 2019), which was set to be the single agreement with the highest number of signatories in history. Countries in the Global North and the Global South managed to reconfirm at COP26 in 2021 their commitment to climate and sustainability goals, which unavoidably may change over time due to geopolitics, fuel reserves and prices, user preferences, as well as technological advancements.

Transport has been consistently one of the sectors with the largest impact on climate as it '*emits around 23% of the energy related CO_2 that feeds global warming*' (ITF, 2021). OECD countries have been the most polluting countries until 2010, although greenhouse gas (GHG) emissions have been growing significantly also in Asian countries since 2010 (Fig. 1.1), mainly due to the growing use of 2-wheelers and 3-wheelers (Sims et al., 2014). If no action is taken, CO_2 emissions by the transport sector could reach 40% by 2030 and increase by 60% until 2050 (ITF, 2021). Hence,

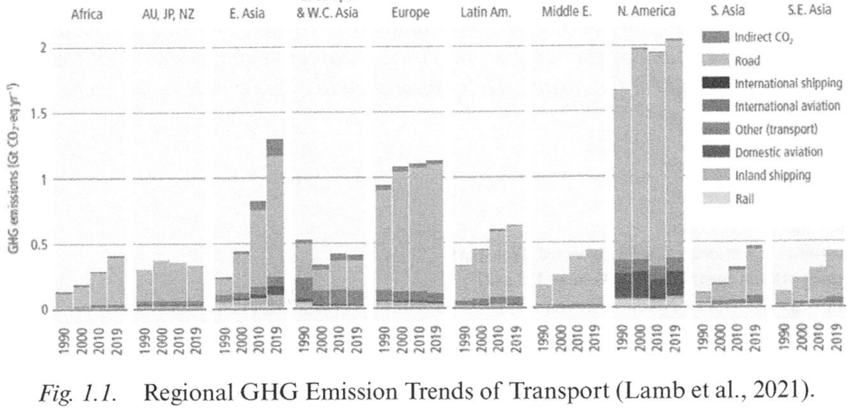

Fig. 1.1. Regional GHG Emission Trends of Transport (Lamb et al., 2021).

a lot of attention and investment have been increasingly directed to addressing the impact of transport worldwide, at least nominally. ACT features as a promising technology-based option to aid in meeting relevant sustainability goals. Technology is therefore an important component in this quest as demonstrated by the potential of, for example, Artificial Intelligence (AI), although it is not and should not be the sole means to address these challenges. Indeed, future transport solutions should move beyond a plain 'techno-fix' approach (Thomopoulos & Givoni, 2015). Transport is included in SDG 11 Sustainable Cities and Communities, although is implicitly contained in other SDGs too:

- SDG 3: Good Health and Well-being
- SDG 5: Gender Equality
- SDG 7: Affordable and Clean Energy
- SDG 8: Decent Work and Economic Growth
- SDG 9: Industry, Innovation and Infrastructure
- SDG 10: Reduced Inequalities
- SDG 11: Sustainable Cities and Communities
- SDG 12: Responsible Consumption and Production
- SDG 13: Climate Action
- SDG 17: Partnerships for the Goals

In assessing the 2022 SDG progress chart (UN, 2022) it is evident that whilst progress on some goals is significant in the developed world and some other parts of the Global South, the overall progress is relatively low (Fig. 1.2). The global perspective is relatively 'fair' to 'limited' progress when it comes to the above SDGs. ICSU-ISSC (2015) stressed that SDG-11 targets are not sufficiently specified and better integration is needed. This is rather alarming as despite the progress made on some sustainability challenges, there is still limited progress about a range of SDGs, particularly regarding the social pillar of sustainability which is primarily reflected on the user perspective (Shiftan et al., 2020). Similar challenges exist regarding the two other sustainability pillars of environmental and economic sustainability, even if some progress has been observed across countries and continents. Undoubtedly, there are other frameworks and indicators to monitor global progress in meeting SDGs, but SDGs are used in this chapter and throughout this book due to their widespread adoption following the 2015 Paris Climate Agreement.

Consequently, this chapter sets the scene for this edited volume first by contrasting ACT deployment with SDGs and then by highlighting the requirement to focus more on addressing user needs through ACT. Due to the growing concerns by scholars about the manner and pace of the transition to ACT, it is pertinent to highlight relevant opportunities and risks. Chapter recommendations foster the promotion of a Living Lab approach based on the Quadruple Helix, which expands the standard Triple Helix model of innovation (Carayannis & Campbell, 2010), to operationalise further the inclusion of social concerns (e.g. gender balance and equity) in policies and actions to meet SDGs across the world.

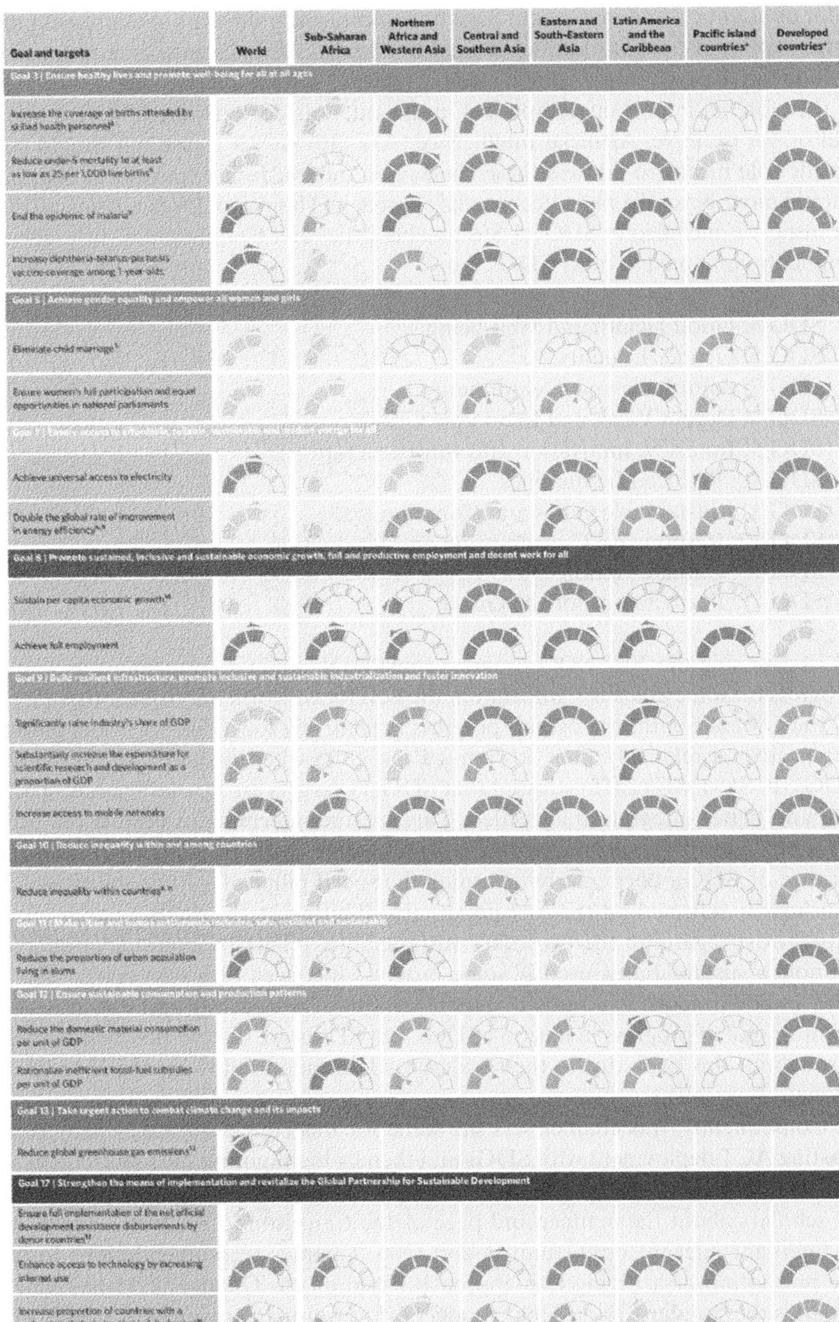

Fig. 1.2. Progress Chart on Selected SDGs (Adapted from UN, 2022) – For the full figure and the latest updates visit: https://unstats.un.org/sdgs/report/2022/progress-chart/

ACT CONTEXT

CO_2 emissions from road transport have increased by more than 37% between 2000 and 2021 (Fig. 1.3), constituting the most polluting transport sub-sector. Based on 2018 rankings by WISE-ACT[1] experts, ACT will introduce challenges and offer solutions primarily for road transport in urban settings (see Fig. 1.4), thus attention should first focus on these modes and settings.

Wider Impacts and Scenario Evaluation of Autonomous & Connected Transport (WISE-ACT) was an Action funded by the COST[2] Programme between 2017 and 2022, which brought together around 200 experts from 42 countries to discuss issues related to ACT implementation.

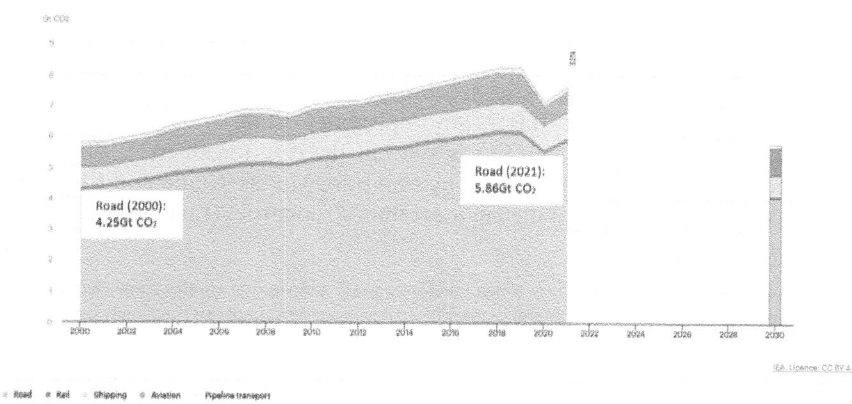

Fig. 1.3. Global CO_2 Emissions from Transport by Sub-Sector in the Net Zero Scenario 2000–2030 (IEA, 2022; Licence: CC BY 4.0).

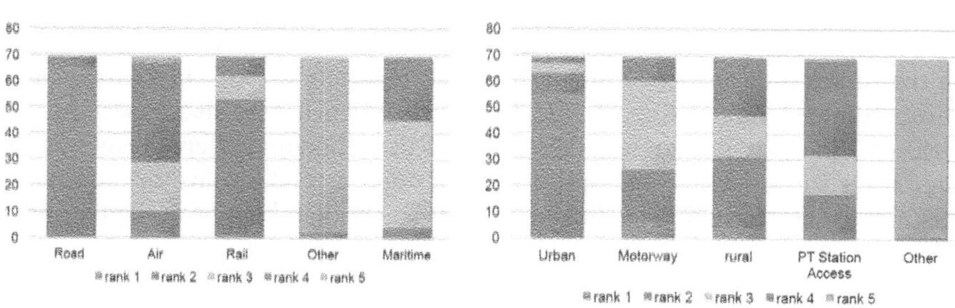

(a) Primary focus of ACT scenarios (b) Setting of primary focus of ACT scenarios

Fig. 1.4. Rankings by WISE-ACT Experts ($n = 35$) Regarding Transport Modes and Settings. (a) Primary Focus of ACT Scenarios. (b) Setting of the Primary Focus of ACT Scenarios. *Source:* Authors' own workings.

[1]WISE-ACT (2017-2022): https://www.wise-act.eu.

[2]European Co-Operation in Science and Technology: https://www.cost.eu.

It is crucial to define ACT and its context before delving into ongoing debates and analyses since there are different definitions used across countries and disciplines. Several terms and acronyms have been used when referring to ACT, for example autonomous, driverless, self-driving cars, automated driving, automated mobility services, AV, CAV, SAV and CCAM (Bala et al., 2023; Harb et al., 2021; Thomopoulos et al., 2021). In spite of relevant terms and acronyms referring to similar technologies, they contain differences which should be clarified and distinguished when planning targets for long-term SDGs. More organisations have been acknowledging the importance of this point, which is demonstrated by the growing use of relevant Glossaries.[3] An early example of such a Glossary attempting to address this need is the WISE-ACT Glossary, which aims at unifying key contemporary ACT terms to support policymakers and practitioners.

Heated debates have taken place among WISE-ACT experts during events and activities about which ACT term to use, mirroring similar debates reported in the academic literature (Hopkins & Schwanen, 2021). Connectivity has been an additional contested element. It is mostly implied to exist for both Automated and Autonomous Vehicles, although this is still not the case for all ACT settings as explained in Chapter 2 of this volume by Makridis, Mattas, Ciuffo and Kouvelas. The WISE-ACT survey distributed in more than 20 countries (Etzioni et al., 2020), used this definition:

> An Autonomous Vehicle is a vehicle which takes over speed and steering control completely and permanently, on all roads and in all situations. The driver-passenger cannot drive manually because the vehicle does not have a steering wheel. The driver-passenger only sets the travel destination.

Lavieri and Bhat (2019) have used a similar one, defining Autonomous Vehicles as:

> Self-driving vehicles, also sometimes referred to as autonomous cars or driverless cars, are capable of responding to the environment and navigating without a human driver in the vehicle controlling the vehicle. In the following questions, whenever you read the term self-driving vehicle, imagine a car with no steering wheel that operates like a personal chauffer.

Whereas, Velasco et al. (2019) defined an Automated Vehicle as 'a vehicle that takes over tasks of human drivers to a certain extent'. In contrast, the Eurobarometer Survey Report (EC-JRC, 2020), which was also based on the WISE-ACT survey, did not use a specific ACT definition and offered respondents three futuristic vehicle image options to consider: a public transport shuttle, a personal car and a truck. 71% of respondents stated that a public transport shuttle corresponds well with their idea of ACT, whereas 60% stated that a personal car also corresponds well with their idea of ACT. Not surprisingly, only 27% of respondents stated that a truck corresponds well with their idea of ACT.

---

[3]For example:
ARCADE (2018–2022) https://www.connectedautomateddriving.eu/glossary/
ELTIS-SUMP (2019) https://www.eltis.org/mobility-plans/glossary
HeadStart (2019–2022) https://www.headstart-project.eu/homepage/glossary
WISE-ACT (2017–2022) https://www.wise-act.eu.

Other scholars have adopted definitions which may be broader or narrow-focused, depending on whether they were primarily targeting ACT technology or ACT users. For example, Milakis (2019) has used this definition:

> Automated driving technology involves hardware (e.g. sensors) and software (e.g. trajectory planning) systems that can assist the driver to conduct the dynamic tasks of driving (e.g. monitor the driving environment, lateral and/or longitudinal motion control).

A lot of scholars do not suggest a specific ACT definition in their studies or projects, whilst others (e.g. Bansal & Kockelman, 2017; Hwang et al., 2021; Kalra & Paddock, 2016) implicitly adopt the definitions issued by SAE (Society of Automotive Engineers) International (SAE J3016) which are also adopted by the International Standards Organisation (ISO, 2021):

> 3.2 AUTOMATED DRIVING SYSTEM (ADS): The hardware and software that are collectively capable of performing the entire DDT on a sustained basis, regardless of whether it is limited to a specific operational design domain (ODD); this term is used specifically to describe a Level 3, 4, or 5 driving automation system.
>
> 3.5 DRIVING AUTOMATION: The performance by hardware/software systems of part or all of the DDT (Dynamic Driving Task) on a sustained basis.
>
> 3.6 DRIVING AUTOMATION SYSTEM OR TECHNOLOGY: The hardware and software that are collectively capable of performing part or all of the DDT on a sustained basis; this term is used generically to describe any system capable of Level 1 to 5 driving automation.
>
> 3.9 DRIVERLESS OPERATION [OF AN ADS-EQUIPPED VEHICLE]: On-road operation of an ADS-equipped vehicle that is unoccupied, or in which on-board users are not drivers or in-vehicle fallback-ready users.

This volume aims to raise awareness about the significance of defining ACT within the context of each specific study or project, particularly given that connectivity is not explicitly included in the SAE definition. Such an approach will facilitate a more holistic evaluation of the existing and projected advantages and disadvantages by practitioners and policymakers.

ACT OPPORTUNITIES AND RISKS

Irrespective of the ACT definition adopted, there has been an increasing volume of academic and 'grey' literature exploring its opportunities and risks, along with ACT acceptance levels (Gkartzonikas et al., 2023; Harb et al., 2021). A review of contemporary literature is available through the five WISE-ACT Working Group Thematic Reports.[4] This section, therefore, does not aim at replicating existing literature but aims to offer a concise summary of the key points about ACT opportunities and risks. This will facilitate comprehension and will provide essential context for the remaining volume chapters.

---

[4]WISE-ACT WG Thematic Reports: https://wise-act.eu/reports-publications/#1592817777293-0a924419-15b2.

More transport modes are increasingly being fitted or outright designed to have automated driving. These include transit and public transport modes i.e. metro, rail and tram services such as in Tampere, Finland. In urban settings where these transport modes can be cost or time-competitive to private cars, they may contribute to the required disruptive changes needed to meet relevant SDGs and wider decarbonisation objectives. At a vehicle level, automation is linked with more efficient driving or eco-driving, which could also contribute to reducing air pollution and congestion levels in cities. That would be particularly helpful for trucks and vans in both urban and inter-urban settings such as motorways. Drones are another promising ACT mode aiming to reduce fuel use and GHG emissions for example, in the delivery of small items such as medical goods. This technology however is still in an embryonic stage. At an individual user level, full automation (e.g. the SAE Level 5) could allow vulnerable road users or those without the ability to drive at a given life stage, access to a private or shared vehicle, thus increasing their accessibility. If shared ACT is adopted by large proportions of users, there is a potential to contribute to climate goals through improved route planning (IPCC, 2022, p. 1256). Such benefits, particularly about shared ACT and equity or inclusion challenges for vulnerable road users, have been outlined in the academic literature (e.g. De Paepe et al., 2023; Etzioni et al., 2020; Kyriakidis et al., 2020) and are coming to the fore through the rise of user-centric approaches in transport planning and policy.[5]

The use of travel time for more productive activities while on ACT has been reported (e.g. Hamadneh & Esztergár-Kiss, 2021) and therefore has the potential to improve the Value of Travel Time (VOT) when compared to driving. Additional opportunities arising from the deployment of ACT may be experienced in tourism and leisure travel (Thomopoulos & Attard, 2022), although it is still unclear what their environmental impact will be as it depends on (i) the level of automation, (ii) if a shared ACT model is adopted and (iii) whether it attracts travellers who previously used cars and taxis, active modes or public transport. Past studies on the impact of transport automation on GHGs found that contemporary projections have been inconclusive, but have the potential to half harmful GHGs particularly when partially automated vehicles are deployed (Wadud et al., 2016) or if combined with Electric Vehicle (EV) technology (Annema, 2020; Nikitas et al., 2021). Sustainable development implications depend on the ACT business models to be adopted and the travellers' behaviour response. These are fields where further research is required (Cabanelas et al., 2023; Stocker & Shaheen, 2017).

On the other hand, a number of ACT risks have been identified and require further analysis to better understand their implications for users, cities,

---

[5]Horizon Europe: https://cordis.europa.eu/programme/id/HORIZON_HORIZON-CL5-2021-D6-01-05.
Horizon Europe: https://ec.europa.eu/info/funding-tenders/opportunities/portal/screen/opportunities/topic-details/horizon-cl5-2023-d6-01-04.

businesses, governments and society at large. Well-documented risks include the so-called 'rebound effect', namely making private mobility more accessible even for currently marginalised groups, which would in turn increase the Vehicle-Kilometres-Travelled (VKT) and reduce the number of public transport users (IPCC, 2022). The release of new AI features in 2023 has renewed past debates about the social impact of ACT and the corresponding ethical challenges (Costantini et al., 2021; Dogan et al., 2020; Martens et al., 2021). Acknowledging such risks, the European Commission appointed an Expert Group to explore these risks in more detail, which were grouped into three high-level areas: data ethics, responsibility and road safety (EC, 2020). Additional issues have been identified by participants and external experts during activities and discussions of this Expert Group, for example, responsible innovation, privacy and data management. These have led to the suggestion to link philosophy and ethics with transport automation, and establish a new field to support the development of ACT (Milakis et al., 2020; Santoni de Sio, 2021), as well as the broader use of responsible innovation frameworks by practitioners and policymakers (Stilgoe et al., 2013).

Along the same lines, Hopkins and Schwanen (2021) have focused on the SAE Levels of Automation (J3016) to highlight the unintended consequences of the widespread use of these SAE levels without a thorough review of their scope and objectives. The SAE Levels of Automation were originally developed to create a universal standard regarding the automation level of a specific AV. Yet, SAE Levels of automation have been widely used to also refer to broader implications beyond a single vehicle on, e.g. a city or a motorway. To address this, Hopkins and Schwanen (2021) focus on both passenger and goods ACT, to provide a useful overview of the potential misuse of the SAE Levels of Automation. A key finding is that the SAE Levels of Automation are largely based on the anglophone world, which unavoidably excludes a lot of other societies, markets and parts of the world. The United States and the United Kingdom aim to be global leaders in ACT, but other countries have also started to develop their own standards (see, e.g. BaST in Germany). Unintended effects of using the SAE Levels of Automation without adjusting them to the context of each study or project were summarised as follows (Hopkins & Schwanen, 2021, p. 5):

- *Illustrating the development of ACT in a linear way overlooks HMI (Human Machine Interaction),*
- *These six levels of automation are simple to meet their primary objective, so definitions and language used are important to address for example enhanced user capabilities and connectivity requirements,*
- *The evaluation of wider impact such as on societies, cities and regions is excluded, so uptake and deployment barriers are not defined to be addressed, and*
- *Multi-modal transport is not considered as the SAE Levels of Automation implicitly focus on the contemporary car-dominated status quo on roads.*

Further to the point made earlier in Section 2, 'ACT Context', about the need to define ACT within each study or project, scholars are increasingly focusing on the need to use existing frameworks and standards appropriately. The 2014 SAE Levels of Automation unavoidably had a core focus on technology and automation. Yet, they lacked the essential conditionality about who, where and how would benefit from ACT (Stayton & Stilgoe, 2020). Therefore, it is pertinent to shift the focus to the various types of ACT users. Without considering and eventually addressing wider and unintended effects, there is a risk of embedding an unjustified optimism about an ACT future. Such a risk cannot be overlooked and it is rather encouraging to see some governments starting to realise this need and planning relevant initiatives. Different approaches have been reported for example in the UK and Finland, since the Finnish approach has been promoting bundled mobility services (e.g. Mobility-as-a-Service MaaS) for more than a decade. Nonetheless, all such initiatives are founded on the underlying principle that: 'if left unaddressed, such lack of reflectivity will not just lead to a plethora of undesired implications for Finnish society at large but will also signify a failure in developing an adaptive governance culture needed to face challenges of the 21st century' (Olin & Mladenovic, 2022). Consequently, it is important to comprehend that there is no single ACT future for every city or state and that there is the need to further embed uncertainty about ACT futures in planning practices.

An overview of potential ACT benefits and threats is offered in Tables 1.1 and 1.2, categorised by impact group that includes users, governments and businesses (Backhaus et al., 2019; Thomopoulos & Givoni, 2015).

Table 1.1. Benefits of ACT.

	Benefits
User	• Travel time is not driving time, so disutility decreases and comfort increases. • Travel time may be used for other work/leisure activities. • AVs can allow the integration of culturally diverse users in cities globally. • Transport-related social exclusion may be eliminated.
Government	• Increased safety due to less accidents (mainly in the era when only AVs will be on the roads). • More parking space will become available and it may be used for other purposes by city authorities. • If AVs are eco-friendly, there could be reduced air pollution and lower energy use from the transport sector. • Accessibility can improve for all travelers, including the elderly and disabled.
Business	• Significant business opportunities will arise for automotive manufacturers, particularly for conventional ones which decide to enter this innovative market. • Expanding databases and innovative use of Big Data will allow for the emergence of business opportunities and new business models, creating value for stakeholders. • Logistics and supply chain business will reduce (congestion, and time) costs through eco-driving, better route planning, V2x communication and platooning.

Source: Adapted from Backhaus et al. (2019) and Thomopoulos and Givoni (2015).

Table 1.2. Threats from ACT.

	Threats
User	• High cost of 'smart' infrastructure (V2I) to accommodate AVs. • Local congestion may increase if the aggregate number of journeys increases. • The cost of emerging mobility patterns can lead to social exclusion of certain groups if high. • Identifying and assigning responsibility for car accidents may become fuzzier. • 'Digital Divide' can lead to increased social exclusion. • Better use of travel time may increase travel time e.g. daily commute, resulting in higher aggregate energy demand at local and national level. • Widespread use of AVs can reduce walking and cycling, increasing obesity and negative health impacts. • Unintended consequences will arise such as privacy, surveillance and data management issues linked with ICT for transport or the threat of wireless hacking to gain unauthorised control of AVs.
Government	• The adjustment period when both conventional human-driven and autonomous cars co-exist on roads could impose more car accidents. • Deciding on the optimal route will be a challenge, particularly during extreme events and principles may differ across cities complicating inter-urban journeys. • The emergence of diverse technologies by competing actors may lead to a lack of coordination and common legislation. • Reduced employment demand for drivers and car technicians, increasing government costs for retraining and/or unemployment benefits.
Business	• Better use of travel time may increase travel time to travel through routes with greater journey comfort leading to increased congestion. • The development of competing technologies by diverse actors may lead to inefficient use of resources and the evolution of competing standards internationally. • Vehicular communication network needs high transmission capacity equipment and proper penetration rate to achieve optimal transport performance.

Source: Adapted from Backhaus et al. (2019) and Thomopoulos and Givoni (2015).

SUSTAINABLE AUTOMATED AND CONNECTED TRANSPORT

After introducing contemporary developments, ongoing debates, as well as ACT benefits and threats, it is appropriate to revert to the fundamental focus of this volume, namely if and how ACT can be sustainable and contribute to meeting the SDGs. The Inter-governmental Panel on Climate Change (IPCC) has attempted in 2022 to map the potential synergies and trade-offs between transport and SDGs (Table 1.3). This is a valuable exercise given the contribution of transport, particularly of road transport, to GHGs.

Yet, existing IPCC reports do not specifically focus on ACT, which seems a significant contemporary gap for academics, practitioners and policymakers. A broad range of new mobility services such as EVs, MaaS or ride-sharing may contribute to meeting SDGs (Cabanelas et al., 2023), but ACT is among the new mobility service sectors which have attracted the highest interest and funding in the 21st century (Milakis et al., 2020). Combining ACT with other new mobility services can provide a unique offering, for example, SAEV (Shared Automated

Table 1.3. Synergies and Trade-Offs Between Transport and SDGs (IPCC, 2022: 1247).

Sustainable Development Goals (SDGs)	Synergies and Trade-Offs
Basic human needs (SDG1, SDG2, SDG3)	Lower air pollution contributes to positive health outcomes Energy access can contribute to poverty alleviation Transport planning is a major player in reducing poverty in cities Access to healthcare Diseases from air pollution Injuries and deaths from traffic accidents Reduced driving-induced stress Links between active transport and good health with positive effects of walking and cycling Improving road accessibility to disabled user Reduce time spent on transport/mobility
Earth preconditions (SDG13, SDG14, SDG15)	Reduction of GHG emissions along the entire value chain, e.g. Well-to-Wheel Further development addressing minor GHG emissions and pollutants Transport-oriented to sustainable development The circular economy principle applied to transport
Sustainable resource use (SDG6, SDG7, SDG12)	Share of renewable energy use Energy efficiency of vehicles Clean and affordable energy off-grid Reduce material consumption during production, lifecycle analysis of vehicles and their operations including value chains Closed loop carbon and nutrient cycle linked to circular economy
Social and economic development (SDG8, SDG9, SDG11)	Role of transport for economic and human development Decarbonised public transport rather than private vehicle use Transport-oriented to sustainable development Sustainable transport infrastructure and system for cities and rural areas Affordability of mobility services, can also be covered under 'universal access' to public transport Accessibility vs mobility; mobility to opportunities; transport equity; development as freedom Positive economic growth (employment) outcomes due to resource efficiency and lower productive energy cost Role of transport provision in accessing work, reconfiguration of social norms, working from home Transport manufacturers as key employers changing role of transport-related labour due to platform economy, innovation in Autonomous Vehicles
University values (SDG4, SDG5, SDG10, SDG17)	Gender equality in transport Reduced inequalities Enables access to quality education Partnership for the goals

Electric Vehicles), as has already been suggested in the literature (for example Nikitas et al., 2021). As mentioned previously ACT can positively contribute to reducing transport accidents through improved tactical safety (Schoener, 2022) and the use of Advanced Driver Assistance Systems (ADAS). This would contribute towards SDG3 and partially towards SDG10. Hence, Table 1.4 is an attempt to map potential ACT contributions in meeting SDGs.

Building on Table 1.4, it is imperative to assess contemporary activities and financial support offered to align technological development and contribute towards meeting the SDGs. Table 1.5 provides an indicative list of 'green vehicle' projects totalling €628 million of European Union funding which also incorporate an ACT element.

For these potentials of ACT to contribute to the SDGs and a more sustainable mobility system, it is important to make good progress not only on the technological front, for example, testing vehicles up to SAE Level of Automation 5, but also on the regulatory and user fronts. It is thus encouraging to see that user-centric approaches are being increasingly introduced and considered. Nevertheless, a framework is needed to ensure that different types of user groups are included from the early stages of the design process. Applying the universal design principles in ACT is a promising approach (Martens et al., 2021). This is currently being tested

Table 1.4. Potential ACT Contributions Towards SDGs (AustriaTech, 2021).

Sustainable Development Goals	Potentials of ACT to Contribute to This Goal
SDG 3 Good health and well-being	Reduce the frequency of accidents and thus fewer traffic fatalities and injuries Consideration and protection of active and passive road users
SDG 9 Industry, innovation and infrastructure	New opportunities and innovation potential through intelligent infrastructure and digital transport structure Optimized utilization and efficient use of existing capacities and solving problems caused by exhausted capacity (e.g., shortage of drivers in freight transport)
SDG 11 Sustainable cities and communities	Improved and more inclusive access to mobility and new forms of public transport In the long term, automated mobility creates the potential for reducing traffic areas and making public space available for other purposes. Higher flexibility also in rural areas (e.g. flexible timetables)
SDG 13 Climate action	'Green Driving', is an efficient driving style that reduces consumption and emissions Improvement of public transport and thereby increasing the acceptance of public transport Interplay of automated mobility with other measures such as alternative propulsion technologies to achieve the Climate Targets of the Paris Agreement

Table 1.5. Green Vehicle H2020 Research Projects Addressing a User Perspective.

Project Start Year	Projects
2018	• SYS2WHEEL[6] • EVC1000[7] • LISA[8]
2019	• USER-CHI[9] • INCIT-EV[10] • eCharge4Drivers[11] • SOLUTIONSplus[12]
2020	• URBANIZED[13] • Multi-Moby[14] • DREEM[15] • LEONARDO[16] • REFLECTIVE[17] • FUTURE-HORIZON[18]

in the INDIMO[19] project for example. Fit-for-purpose Key Performance Indicators (KPIs) need to be developed as part of emerging frameworks to address all three pillars of sustainability, namely economic, environmental and social.

Expanding the Triple Helix innovation model (Etzkowitz & Leyedesdorff, 2000), Carayannis and Campbell (2010) introduced the Quadruple and eventually the Quintuple Helix innovation models to meet the SDGs. The Quadruple Helix has been linked with a Living Lab approach in the EU-MACS[20] project. These frameworks have been only recently introduced in the transport field, but are still absent from ACT research. Nonetheless, they seem as a germane component for a framework evaluating how ACT can meet SDGs. The novelty of the Quadruple Helix has been the introduction of users and civic society organisations to

[6]https://cordis.europa.eu/project/id/824244
[7]https://cordis.europa.eu/project/id/824250
[8]https://cordis.europa.eu/project/id/814471
[9]https://cordis.europa.eu/project/id/875187
[10]https://cordis.europa.eu/project/id/875683
[11]https://cordis.europa.eu/project/id/875131
[12]https://cordis.europa.eu/project/id/875041
[13]https://cordis.europa.eu/project/id/101006943
[14]https://cordis.europa.eu/project/id/101006953
[15]https://cordis.europa.eu/project/id/101007085
[16]https://cordis.europa.eu/project/id/101006687
[17]https://cordis.europa.eu/project/id/101006747
[18]https://cordis.europa.eu/project/id/101006598
[19]INDIMO: https://www.indimoproject.eu.
[20]EU-MACS: https://eu-macs.eu/outputs/livinglabs/.

the already existing three-dimensional model of Academia – Government – Industry. Moreover, the Quintuple Helix has added the natural environment as a fifth dimension. Undoubtedly, there are valid debates still taking place on whether the additional helix dimensions should constitute new dimensions or just components of the previous three or four helices. Interestingly, both the fourth and fifth helices seem indispensable components of ACT development, if ACT is to become an innovation fulfilling the SDGs. The user and societal perspective are essential for ACT deployment (Shiftan et al., 2020) and the natural environment also has a core role linked with ACT trials.

As a result, this chapter's recommendations foster the promotion of a Quintuple Helix approach to operationalise the inclusion of not only environmental (IPCC, 2022) but also social and economic concerns of all relevant stakeholders in SUMP across the world. Living Labs are one of the suggested approaches to achieve this since they revolve around a user-centred co-creative approach, which can increase support or highlight deficiencies during ACT development (Kaplan et al., 2023). A number of shortcomings have started being reported given the growth of Living Labs as an innovative approach, mainly due to the lack of systematic integration of practical outputs (Bylund et al., 2022). The same has been observed with ACT during the first decade of innovation in this field and the public testing of higher levels of automation through an increasing number of ACT trials and Living Labs. The Connected Cooperative Automated Mobility (CCAM) Association, for example, has acknowledged this and has included a dedicated cluster about societal aspects and user needs. To conclude, it should be clear that ACT is a very promising innovation, but its full potential can only be realised and shared across all stakeholders if SDGs are incorporated into a holistic evaluation framework. Only then responsible innovation in relation to anticipatory, participatory and institutional aspects of governance will be feasible (Genus & Stirling, 2018).

OVERVIEW OF THIS VOLUME

To address the contemporary challenges and ensure that ACT contributes to the SDGs, this book is divided into two parts. The first part presents selected ACT innovations to illustrate the potential benefits of ACT and their anticipated contribution towards SDGs. The second part presents a number of policy and regulatory issues surrounding the development and deployment of ACT, with a primary focus on Europe.

The analysis by Makridis, Mattas, Ciuffo and Kouvelas focuses on existing lower automation levels offered through Adaptive Cruise Control (ACC) and Cooperative Adaptive Cruise Control (CACC) to illustrate the impact of connectivity and platooning on traffic safety, driving behaviour, and energy consumption. Through their traffic simulations, they demonstrate that CACC holds stronger potential to contribute to meeting SDGs. Yet, the energy requirements to move from existing ACC to widely adopted CACC technologies have not been studied adequately to date, so further research is required about it.

Expanding the focus on simulation studies, Dadashzadeh, Agriesti, Sadid, Elvarsson, Roncoli and Antoniou provide an overview of existing approaches within the ACT domain. An interesting element is the treatment of uncertainty within the very diverse simulation scenarios and the variables used. Hence, a number of KPIs are proposed, which can be used to develop ACT simulation scenarios and test specific features related to SDGs.

An innovative setting is outlined by Etzioni, Collins, Ben-Elia, and Shiftan using a Serious Games approach. Three novel automated choice options are offered to serious game players: (i) shared ride, (ii) shared car and (iii) automated transit (i.e. public transport), to improve player learning through visualisation, feedback and game scores. Criteria used by players to improve their game score include mode choice, travel punctuality, and the choices of other players. Additional features to be explored in the future could focus on privacy, ride-sharing with strangers, and unpleasant odours as has been suggested in the literature (see Thomopoulos et al., 2021). Their future research suggestion to use Virtual Reality is truly innovative and could provide an inclusive and immersive user-centric option for ACT development.

The second part of this book is introduced by Nikitas, who addresses the rhetorical question of whether driverless transport can be sustainable. Using a Triple Helix-based, thematically organised narrative review, this chapter lays the foundation for sustainable ACT, acting as a bridge with Section 4, 'Sustainable Automated and Connected Transport' of Chapter 1. Links with Chapter 1 are evident since Nikitas argues that a techno-fix approach is not sufficient, if not coupled with a far-reaching travel behaviour change. He concludes that educating stakeholders through Living Labs would be beneficial. He argues for combining ACT with alternative fuelling, multi-modality and increased use of public transport.

Regulatory frameworks to support ACT testing and deployment are still rare, although there is increasing appetite by both policymakers and the industry. Shallauer, Soteropoulos, Cornet, Klar and Fürdös offer a comparative analysis of ACT national regulations, linking them with their implementation to meet wider SDGs. Reviewing national laws, acts, ordinances and decrees as well as using evidence from the H2020 SHOW project, they find a high level of disparity between European countries. They recommend stronger cross-border coordination and an increased focus on user needs, to ensure that national and local SDGs will be met.

Gavanas outlines the possible impacts of autonomous vehicles and discusses their uncertainty and trade-offs in relation to sustainable urban development and planning. It links the implementation of ACT to SDG11 goals and discusses the main challenges for social and economic sustainability. It highlights the potential of data and big data from such technologies in planning and recommends a data governance framework for ACT deployment. It draws up recommendations for planners and concludes on the key issues surrounding ACT.

Beyazit, Soh and Martens discuss the social challenges of ACT by focusing on how it can reduce inequities and make the transport system more inclusive. This chapter reflects on four distinct ACT deployment scenarios by assessing their opportunities and threats. Employing Singapore as a case study, it allows the authors to draw governance and policy-making conclusions, which highlight

ACT-based public transport as the most promising option for an inclusive mobility future.

A critical element of any innovation model which aims at a long-term future is the business model it is built upon. Consequently, Parkhurst, Cabanelas and Paddeu provide a comprehensive overview of selected technological advancements and their potential business model implications across transport modes for both passenger and freight transport. By taking a critical view towards SDGs whilst accepting their *de facto* dominance among practitioners and policymakers, they provide an innovative framework to appraise viable business models for sustainable mobility policies. Interestingly, they conclude that Hydrogen Heavy Goods Vehicles and Electric Cars will only have a limited positive impact on meeting key SDGs. Furthermore, they include employment and workforce skills impacts in their analysis, which provides additional value for future impact assessment through a review of capabilities and capacities.

Overall, this volume addresses key considerations about Sustainable ACT in order to provide a valuable resource of evidence-based ACT information for academics, practitioners and policy-makers in various countries. Its contribution is in providing a holistic overview of methods and frameworks to inform ACT deployment and the development of relevant regulations that contribute to meeting wider societal objectives and SDGs.

ACKNOWLEDGEMENTS

The authors are thankful for the support provided by the WISE-ACT CA16222 COST Action through workshops, Short-Term Scientific Missions, Training Schools, events and conferences, which contributed to forming collaborations and generating input for this book. We would also like to thank the authors and reviewers of each chapter for their time and effort in contributing to this volume.

REFERENCES

Annema, J. A. (2020). Policy implications of the potential carbon dioxide (CO_2) emission and energy impacts of highly automated vehicles. In D. Milakis, N. Thomopoulos, & B. Van Wee (Eds.), *Policy implications of autonomous vehicles, Advances in Transport Policy and Planning Series* (Vol. 5, pp. 149–162). Academic Press – Elsevier.

AustriaTech. (2021). *Automated Mobility in Austria – Monitoring Report 2020 (Automatisierte Mobilität in Österreich – Monitoringbericht 2020)*. https://www.austriatech.at/assets/Uploads/Publikationen/PDF-Dateien/30033972bc/Automatisierte_Mobilitat_in_Oesterreich_2020.pdf

Backhaus, W., Rupprecht, S., Franco, D., Hoadley, S., van den Bosch, F., Ivari, M., Matthews, B., Anwar, A., Miles, J., Gaietto, G., Flechon, C., Sukennik, P., Dahl, A., Faul, N., Goger, T., Olstam, J., Friedrich, M., Alessandrini, A., & Thomopoulos, N. (2019). *Road vehicle automation in sustainable urban mobility planning – practitioners briefing, ELTIS and CoEXist H2020 Grant Agreement 723201*. Rupprecht Consulting.

Bala, H., Anowar, S., Chng, S., & Cheah, L. (2023). Review of studies on public acceptability and acceptance of shared autonomous mobility services: Past, present and future. *Transport Reviews*, 45(5), 970–996, https://doi.org/10.1080/01441647.2023.2188619

Bansal, P., & Kockelman, K. M. (2017). Forecasting Americans' long-term adoption of connected and autonomous vehicle technologies. *Transportation Research – Part A: Policy and Practice*, *95*, 49–63. https://doi.org/10.1016/j.tra.2016.10.013

Bylund, J., Riegler, J., & Wrangsten, C. (2022). Anticipating experimentation as the 'new normal' through urban living labs 2.0: Lessons learnt by JPI Urban Europe. *Urban Transformations*, *4*(8). https://doi.org/10.1186/s42854-022-00037-5

Cabanelas, P., Parkhurst, G., Thomopoulos, N., & Lampón, J. F. (2023). A dynamic capability evaluation of emerging business models for new mobility. *Research in Transportation Business & Management*, *47*, 100964. https://doi.org/10.1016/j.rtbm.2023.100964

Carayannis, E., & Campbell, D. (2010). Triple helix, quadruple helix and quintuple helix and how do knowledge, innovation and the environment relate to each other? A proposed framework for a trans-disciplinary analysis of sustainable development and social ecology. *International Journal of Social Ecology and Sustainable Development*, *1*(1), 41–69. https://doi.org/10.4018/jsesd.2010010l0

Climate Change. (2022). Mitigation of Climate Change. In P. R. Shukla, J. Skea, R. Slade, A. Al Khourdajie, R. van Diemen, D. McCollum, M. Pathak, S. Some, P. Vyas, R. Fradera, M. Belkacemi, A. Hasija, G. Lisboa, S. Luz & J. Malley (Eds.), *Contribution of Working Group III to the Sixth Assessment Report of the Intergovernmental Panel on Climate Change*. Cambridge University Press.

Costantini, F., di Ciommo, F., de Jesus Mateo Sanguino, T., Carballo-Calero, P., Klar, W., Shallauer, D., Nguyen-Luong, D., Dogan, E., Arora, P., Backhaus, W., Franco, D., Rupprecht, S., Susilo, Y., Shiko, V., & Thomopoulos, N. (2021). *Institutional and Regulatory Challenges of Autonomous and Connected Transport*. WG1 Thematic Report, COST Action CA16222, WISE-ACT.

De Paepe, L., Van Acker, V., & Witlox, F. (2023). To share or not to share, by whom is the question. Acceptability and acceptance of shared transport services by vulnerable groups. *Transport Reviews*, *43*(5), 935–969. https://doi.org/10.1080/01441647.2023.2185314

Dogan, E., Costantini, F., & Le Boennec, R. (2020). Ethical issues concerning automated vehicles and their implications for transport. In D. Milakis, N. Thomopoulos, & B. van Wee (Eds.), *Policy implications of autonomous vehicles, Advances in Transport Policy and Planning Series* (Vol. 5, pp. 215–233). Academic Press – Elsevier.

European Commission (EC). (2020). *Ethics of connected and automated vehicles: Recommendations on road safety, privacy, fairness, explainability and responsibility, Horizon 2020 Commission Expert Group to advise on specific ethical issues raised by driverless mobility (E03659)*. Publication Office of the European Union. https://research-and-innovation.ec.europa.eu/news/all-research-and-innovation-news/new-recommendations-safe-and-ethical-transition-towards-driverless-mobility-2020-09-18_en

European Commission – Joint Research Centre (EC-JRC). (2020). *Expectations and concerns of connected and automated driving, Special Eurobarometer 496 Report, Wave EB92.1, Joint Research Centre*. European Commission. https://europa.eu/eurobarometer/surveys/detail/2231

European Green Vehicles Initiative (EGVI). (2021). *Project Portfolio Calls 2018-2020, Horizon 2020 European Union Funding for Innovation and Research*. European Green Vehicles Initiative.

Etzioni, S., Hamadneh, J., Elvarsson, A., Esztergár-Kiss, D., Djukanovic, M., Neophytou, S., Sodnik, J., Polydoropoulou, A., Tsouros, I., Pronello, C., Thomopoulos, N., & Shiftan, Y. (2020). Modelling cross-national differences in Automated Vehicle acceptance. *Sustainability*, *12*, 9765. https://doi.org/10.3390/su12229765

Etzkowitz, H., & Leyedesdorff, L. (2000). The dynamics of innovation: From National Systems and "Mode 2" to a Triple Helix of university-industry-government relations. *Research Policy*, *29*(2), 109–123. https://doi.org/10.1016/S0048-7333(99)00055-4

Genus, A., & Stirling, A. (2018). Collingridge and the dilemma of control: Towards responsible and accountable innovation. *Research Policy*, *47*(1), 61–69. https://doi.org/10.1016/j.respol.2017.09.012

Gkartzonikas, C., Losada-Rojas, L. L., Christ, S., Pyrialakou, V. D., & Gkritza, K. (2023). A multi-group analysis of the behavioral intention to ride in autonomous vehicles: Evidence from three US metropolitan areas. *Transportation*, *50*(2), 635–675. https://doi.org/10.1007/s11116-021-10256-7

Hamadneh, J., & Esztergár-Kiss, D. (2021). The influence of introducing autonomous vehicles on conventional transport modes and travel time. *Energies, 14*(14), 4163. https://doi.org/10.3390/en14144163

Harb, M., Stathopoulos, A., Shiftan, Y., & Walker, J. L. (2021). What do we (not) know about our future with automated vehicles? *Transportation Research Part C: Emerging Technologies, 123*, 102948. https://doi.org/10.1016/j.trc.2020.102948

Hopkins, D., & Schwanen, T. (2021). Talking about automated vehicles: What do levels of automation do? *Technology in Society, 64*, 101488. https://doi.org/10.1016/j.techsoc.2020.101488

Hwang, J., Li, W., Stough, L. M., Lee, C., & Turnbull, K. (2021). People with disabilities' perceptions of autonomous vehicles as a viable transportation option to improve mobility: An exploratory study using mixed methods. *International Journal of Sustainable Transportation, 15*(12), 924–942. https://doi.org/10.1080/15568318.2020.1833115

ICSU-ISSC. (2015). Review of the sustainable development goals: The science perspective. Paris: International Council for Science. https://council.science/publications/review-of-targets-for-the-sustainable-development-goals-the-science-perspective-2015/

International Energy Agency (IEA). (2022). Global CO_2 emissions from transport – Data and Statistics, International Energy Agency. https://www.iea.org/data-and-statistics/charts/global-co2-emissions-from-transport-by-sub-sector-in-the-net-zero-scenario-2000-2030

Intergovernmental Panel on Climate Change (IPCC). (2022). *Climate Change 2022: Mitigation of Climate Change – Working Group III Contribution to the Sixth Assessment Report of the Intergovernmental Panel on Climate Change, WMO – World Meteorological Organization & UNEP – United Nations Environment Programme*. https://www.ipcc.ch/report/sixth-assessment-report-working-group-3/

International Standards Organisation (ISO). (2021). *ISO/SAE PAS 22736:2021(en) Taxonomy and definitions for terms related to driving automation systems for on-road motor vehicles*. https://www.iso.org/obp/ui/#iso:std:iso-sae:pas:22736:ed-1:v1:en

International Transport Forum (ITF). (2021). *Decarbonising Transport Initiative, Transport Outlook, International Transport Forum*. https://www.itf-oecd.org/decarbonising-transport

Kalra, N., & Paddock, S. M. (2016). Driving to safety: How many miles of driving would it take to demonstrate autonomous vehicle reliability? *Transportation Research - Part A: Policy and Practice, 94*, 182–193. https://doi.org/10.1016/j.tra.2016.09.010

Kaplan, L., Rupprecht, S., Grosso, M., Thomopoulos, N., Backhaus, W., Raposo, M., Ciuffo, B., & Franco, D. (2023). Ensuring strong public support for automation in the planning process: From engagement to co-creation. In G. Meyer & S. Beiker (Eds.), *Road Vehicle Automation 9* (pp. 167–183). Springer.

Kyriakidis, M., Sodnik, J., Stojmenova, K., Elvarsson, A., Pronello, C., & Thomopoulos, N. (2020). The role of human operators in safety perception of AV development – Insights from a large European survey. *Sustainability, 12*, 9166. https://doi.org/10.3390/su12219166

Lamb, W., Wiedmann, T., Pongratz, J., Andrew, R., Crippa, M., Olivier, J., Wiedenhofer, D., Mattioli, G., Al Khourdajie, A., House, J., Pachauri, S., Figueroa, M., Saheb, Y., Slade, R., Hubacek., K., Sun, L., Khan Ribeiro S., Khennas, S., de la Rue du Can, S., Chapungu, L., Davis, S., Bashmakov, I., Dai, H., Dhakal, S., Tan, X., Geng, Y., Gu, B., & Minx, J. (2021). A review of trends and drivers of greenhouse gas emissions by sector from 1990 to 2018. *Environental Research Letters, 16*, 073005. DOI: 10.1088/1748-9326/abee4e

Lavieri, P. S., & Bhat, C. R. (2019). Modeling individuals' willingness to share trips with strangers in an autonomous vehicle future. *Transportation Research - Part A: Policy and practice, 124*, 242–261. https://doi.org/10.1016/j.tra.2019.03.009

Martens, K., Beyazit, E., Henenson, E., Thomopoulos, N., Milakis, D., Mladenovic, M., Pudane, B., van Wee, B., Di Ciommo, F., Curl, A., Cugurullo, F., Dimitrova, E., & Negulescu, M. (2021). *Autonomous and Connected Transport as part of an inclusive transport system. WG2 Thematic Report, COST Action CA16222*. WISE-ACT.

Milakis, D. (2019). Long-term implications of automated vehicles: An introduction. *Transport Reviews, 39*(1), 1–8. https://doi.org/10.1080/01441647.2019.1545286

Milakis, D., Thomopoulos, N., & Van Wee, B. (2020). Policy implications of autonomous vehicles, *Advances in Transport Policy and Planning Series*. Academic Press – Elsevier.

Nikitas, A., Thomopoulos, N., & Milakis, D. (2021). The environmental and resource dimensions of automated transport: A nexus for enabling vehicle automation to support sustainable urban mobility. *Annual Review of Environment and Resources*, *46*, 167–192. https://doi.org/10.1146/annurev-environ-012220-024657

Olin, J., & Mladenovic, M. (2022). Imaginaries of road transport automation in Finnish governance culture – A critical discourse analysis. *Sustainability*, *14*, 1437. https://doi.org/10.3390/su14031437

Santoni de Sio, F. (2021). The European Commission report on ethics of Connected and Automated Vehicles and the future of ethics of transportation. *Ethics and Information Technology*, *23*, 713–726. https://doi.org/10.1007/s10676-021-09609-8

Schoener, H. P. (2022). Assuring responsible driving of autonomous vehicles. In 33rd IEEE Intelligent Vehicles Symposium, Aachen, Germany.

Shiftan, Y., Polydoropoulou, A., Thomopoulos, N., & Rappazzo, V. (2020). *Special Issue "Autonomous and Connected Transport – The user perspective"*. Sustainability. https://www.mdpi.com/journal/sustainability/special_issues/Automated_Connected_Transport

Sims, R., Schaeffer, R., Creutzig, F., Cruz-Núñez, X., D'Agosto, M., Dimitriu, D., Figueroa Meza, M. J., Fulton, L., Kobayashi, S., Lah, O., McKinnon, A., Newman, P., Ouyang, M., Schauer, J. J., Sperling, D., & Tiwari, G. (2014). Transport. In O. Edenhofer, R. Pichs-Madruga, Y. Sokona, E. Farahani, S. Kadner, K. Seyboth, A. Adler, I. Baum, S. Brunner, P. Eickemeier, B. Kriemann, J. Savolainen, S. Schlömer, C. von Stechow, T. Zwickel, & J. C. Minx (Eds.), *Climate change: Mitigation of climate change. Contribution of Working Group III to the Fifth Assessment Report of the Intergovernmental Panel on Climate Change*. Cambridge University Press.

Stayton, E., & Stilgoe, J. (2020). It's time to rethink levels of automation for self-driving vehicles. *IEEE Technology and Society Magazine*, *39*(3), 13–19. https://doi.org/10.1109/MTS.2020.3012315

Stilgoe, J., Owen, R., & Macnaghten, P. (2013). Developing a framework for responsible innovation. *Research Policy*, *42*(9), 1568–1580. https://doi.org/10.1016/j.respol.2013.05.008

Stocker, A., & Shaheen, S. (2017). Shared automated vehicles: Review of business models. Discussion Paper. *International Transport Forum*. https://www.itf-oecd.org/sites/default/files/docs/shared-automated-vehicles-business-models.pdf

Thomopoulos, N., & Attard, M. (2022). Autonomous vehicles. In D. Buhalis (Ed.), *Encyclopedia of tourism management and marketing*. Edward Elgar Publishing.

Thomopoulos, N., Cohen, S., Hopkins, D., Siegel, L., & Kimber, S. (2021). All work and no play? Autonomous vehicles and non-commuting journeys. *Transport Reviews*, *41*(4), 456–477. https://doi.org/10.1080/01441647.2020.1857460

Thomopoulos, N., & Givoni, M. (2015). The autonomous car – A blessing or a curse for the future of low carbon mobility? An exploration of likely vs. desirable outcomes. *European Journal of Futures Research*, *3*(1), 1–14. https://doi.org/10.1007/s40309-015-0071-z

United Nations (UN). (2022). *Sustainable development goals progress chart*. https://unstats.un.org/sdgs/report/2022/progress-chart/

UN-IGS. (2019). *Global Sustainable Development Report – The future is now: Science for achieving sustainable development, Independent Group of Scientists appointed by the Secretary –General*. Department of Economic and Social Affairs – United Nations. https://sdgs.un.org/gsdr/gsdr2019

Velasco, J. P. N., Farah, H., van Arem, B., & Hagenzieker, M. P. (2019). Studying pedestrians' crossing behavior when interacting with automated vehicles using virtual reality. *Transportation Research Part F: Traffic Psychology and Behaviour*, *66*, 1–14. https://doi.org/10.1016/j.trf.2019.08.015

Wadud, Z., MacKenzie, D., & Leiby, P. (2016). Help or hindrance? The travel, energy and carbon impacts of highly automated vehicles. *Transportation Research Part A: Policy and Practice*, *86*, 1–18. https://doi.org/10.1016/j.tra.2015.12.001

CHAPTER 2

IMPACTS OF PARTIALLY CONNECTED AND AUTOMATED VEHICLES ON TRAFFIC FLOW AND ENERGY BASED ON WORLDWIDE EXPERIMENTAL OBSERVATIONS IN MOTORWAY DRIVING

Michail A. Makridis[a], Konstantinos Mattas[b], Biagio Ciuffo[b] and Anastasios Kouvelas[a]

[a]*ETH Zürich, Institute for Transport Planning and Systems (IVT), Zürich, Switzerland*
[b]*European Commission – Joint Research Centre, Ispra (VA), Italy*

ABSTRACT

Road transport networks might face the most significant transformation in the following decades, mostly due to the anticipated introduction of Connected and Automated Vehicles (CAVs). The introduction of connectivity and automation will be realised gradually. There are distinctive levels of automation starting from single-dimension automated functionalities, such as regulating the vehicle's longitudinal behaviour via Adaptive Cruise Control (ACC) systems. Although the technological readiness level is undeniably far from full vehicle automation, there are already commercially available lower-level

Sustainable Automated and Connected Transport
Transport and Sustainability, Volume 19, 23–45

ISSN: 2044-9941/doi:10.1108/S2044-994120240000019002

automated vehicles. The penetration rate of vehicles equipped with Advanced Driver Assistance Systems (ADAS) such as ACC or Cooperative-ACC is constantly increasing bringing new driving behaviours into existing infrastructure, especially on motorways. Lately, several experiments have been conducted with platoons of ACC and CACC-equipped vehicles aiming to study the characteristics and properties of the traffic flow composed by them. This chapter aims to gather the most significant efforts on the topic and present the recent status of research and policy. The impact analysis presented within this chapter is multi-dimensional spanning from traffic flow oscillations and string stability, traffic safety to driving behaviour, energy consumption, and policy, all factors where automation has the potential to contribute to a more sustainable transport system. Investigations through analytical approaches and simulation studies are discussed as well, in comparison to empirical insights, attempting to generalise experimental conclusions. At the end of this chapter, the reader should have a clear view of the existing and potential benefits of CAVs but also the existing and future challenges they can bring.

Keywords: Adaptive cruise control; traffic flow; car platoons; intelligent transport systems; string stability; traffic hysteresis; connectivity; cooperative adaptive cruise control

INTRODUCTION

Road transport networks might face the most significant transformation in the following decades, mostly due to the anticipated introduction of Connected and Automated Vehicles (CAVs) towards more sustainable road transport networks. Many policymakers, practitioners, and researchers expect CAVs to decrease traffic congestion and energy consumption, leading to more sustainable road transport. More specifically, CAVs can offer unprecedented traffic management and control strategies to alleviate or completely prevent complex traffic phenomena such as phantom traffic jams, traffic oscillations, instability, capacity drop, traffic hysteresis, and others, while at the same time promoting system optimal solutions with regard to both travel times and energy consumption. Additionally, new driving behaviours have been arising from the intrinsic differences between controllers and human drivers. The first UN Regulation on the approval of automated driving systems was adopted in 2022 (UNECE, 2022). It concerns level 3 highway automated driving systems (referred to as Automated Lane Keeping Systems, ALKS, although in the latest amendment lane changing capabilities are also foreseen), and the first worldwide regulation concerning level 4 urban and peri-urban automated driving systems (in particular focusing on automated shuttles, robotaxis, hub-to-hub vehicles, and valet parking services) has been adopted in the EU (EC, 2022). These two milestones open the market introduction of automated driving systems and will mark a progressive shift from testing to actual commercialisation.

Still, the introduction of connectivity and automation will be neither instantaneous nor simultaneous. There are distinctive levels of automation starting from single-dimension automated functionalities, such as regulating the vehicle's longitudinal behaviour via ACC systems. The higher level of automation corresponds to vehicles able to drive without any human supervision and under any driving, weather, or other conditions. Although the technological readiness level is undeniably far from full vehicle automation, there are already commercially available lower-level automated vehicles. The penetration rate of vehicles equipped with ADAS such as ACC or Cooperative-ACC, Lane Keeping Assist, Collision Avoidance System, Lane Change Assistance, and others is constantly increasing bringing new driving behaviours into existing infrastructure, especially on motorways. Connectivity developments for traffic or energy-related purposes are not deployed yet but different cooperation classes have been defined, and it is argued that connectivity is necessary to achieve optimal results.

The impact of automation and connectivity is of common interest to all involved stakeholders, i.e. the industry, the policymakers, the academia, the public authorities, and even individual buyers. In the literature, there are bold expectations of the benefits that automated driving can bring to road users but most conclusions so far are based on ambiguous assumptions or modelling approximations. Empirical observations of highly automated vehicles could help affirm those assumptions or set more suitable ones if the existing assumptions fail to accurately describe the systems.

Lately, several experiments have been conducted with platoons of ACC and CACC-equipped vehicles aiming to study the characteristics and properties of the traffic flow composed by them. Such datasets pave the way for shedding more light on the topic and understanding the potential for more sustainable future transport. Real-world observations from commercially deployed systems can uniquely support research on the topic. On the other hand, it is impossible to obtain data for all possible scenarios and circumstances. Consequently, analytical investigations based on available empirical datasets can extrapolate and generalise theoretical conclusions with higher certainty. This chapter aims to gather the most significant efforts on the topic and present the status of research and policy. The impact analysis presented within this chapter is multi-dimensional spanning from traffic flow oscillations and string stability, traffic safety to driving behaviour, energy consumption, and policy. Investigations through analytical approaches and simulation studies are discussed as well, in comparison to empirical insights, attempting to generalise experimental conclusions. At the end of this chapter, the reader should have a clear view of the existing and potential benefits of CAVs but also the existing and future challenges they can bring.

EXPERIMENTAL OBSERVATIONS OF (COOPERATIVE-) ADAPTIVE CRUISE CONTROL SYSTEMS

Both ACC systems and ACC with cooperation (CACC) are considered the predecessors of vehicle automation with regard to longitudinal vehicle movement.

The maturity of ACC systems in the market is already high and therefore future fully automated vehicles will, most probably, use as a basis the knowledge obtained by deployed ACC and CACC systems. It has been over 20 years since the first generation of ACC-equipped vehicles were available on the market, but the design of control strategies for the commercially available ACC systems are still closely guarded secrets of their industrial developers. For an informative review of the available design of control strategies for ACC systems implemented on experimental vehicles, we refer the reader to the work by He et al. (2019).

It should be mentioned that the platooning concept originated for trucks and not passenger vehicles. Trucks are driven by professionals continuously for many hours, a tedious driving task that leads to high fuel/energy consumption. An important role in the high energy demand is the opposing force due to air drag. The original platoon concept implies some sort of communication between consequent trucks to keep very short spacing from the leading truck without compromising safety, while at the same time facilitating comfort, minimising opposing air drag forces, and thus reducing consumption for the follower. This platoon concept is not feasible without short-range communication. For passenger vehicles with ACC systems, the platoon concept implies a vehicle driving behind a leader, with a distance corresponding to the ACC setting set by the driver. The latter platoon concept, often called car-platoon, is used throughout this chapter in all different levels of analysis (even with CACC trajectories).

Although precise operational properties of commercial ACC systems are not disclosed, it is clear from available observations that these systems work on a constant headway principle, that is under stable driving conditions for motorway speeds, they follow the leading vehicle with a desired time headway value (usually between 1 and 2 seconds). ACC-enabled vehicles are equipped with (at least one between) radar and camera sensors in order to monitor the distance from- and the speed of the preceding vehicle. CACC systems are equipped additionally with some sort of communication. The communication protocols are not yet market-ready. Possible ways of communication involve vehicle-to-vehicle (V2V), vehicle-to-infrastructure (V2I), or vehicle-to-everything (V2X), implying communication with mobile or static actors, pedestrians, bicycles, etc. We refer the reader to the most recent communication standards published by SAE International (SAE, 2020).

In order to provide quantitative findings on the impact of ACC and CACC systems upon mass deployment in future transport networks, researchers focus on experimental observations involving car-platoons of ACC-equipped vehicles. Lately, an increasing number of experimental campaigns involving such car platoons appear in the literature and some of them are publicly available (Open Access) to researchers for further investigations. Meanwhile, data processing algorithms are being proposed for the specific problem of reconstructed vehicle trajectory data (Makridis & Kouvelas, 2023). A brief introduction of important datasets currently available for further investigations follows. Please note that this list of tests is not extensive as more and more data are becoming available.

OpenACC Dataset

The OpenACC dataset was initiated by the open-access policy of the Joint Research Centre of the European Commission and involves a set of car-following experiments with human-driven and ACC-driven vehicles on public roads, test tracks, and the JRC private urban network. The dataset includes measurements from different vehicle models and brands, and various data acquisition systems, from inexpensive U-blox and OBD devices to high-precision Differential GNSS systems. This is the largest publicly available database with car-platoon experimental observations, it is constantly populated with new measurements and the reader can download it from the website (JRC-EC, 2020). More details about the OpenACC can be found in the work by Makridis et al. (2021).

Vanderbilt University Dataset

The Vanderbilt University dataset includes two campaigns with car-following data of ACC-driven vehicles. During the first campaign, four-speed profiles are recorded to observe the behaviour of each vehicle in the two-vehicle tests. For all tests, the vehicles begin on the track and start at a low speed with a full-size sedan as the lead vehicle and the test vehicle as the follower vehicle. To validate the emergent traffic flow behaviour of ACC vehicles at the aggregate (system) level, a second campaign was conducted where seven identical ACC vehicles follow a lead vehicle that creates a velocity slow-down event. Position and speed data for each vehicle were collected using high-accuracy uBlox EVK-M8T GPS receivers. For more information, we refer the reader to the corresponding publications (Gunter, Gloudemans, et al., 2020; Gunter, Janssen, et al., 2020).

MA and GA Experiments

Extensive field experiments using four different commercial ACC car models (referred to as MA experiments) investigated their car-following behaviours under disturbances (Li et al., 2021). Further experiments focusing on an ACC-equipped vehicle (referred to as GA experiments) were conducted on an urban arterial road with a two-vehicle car-platoon setup. The leader was driven in cruise control at multiple speed levels and the following vehicle was always engaged with ACC. The ACC remained enabled even for near-stop speeds providing direct findings for the jam density of such systems. Both tests used U-blox devices and measurements obtained at 10 Hz (Li et al., 2022). An online initiative by the same authors attempts to gather all available worldwide experiments in a single repository.[1]

CARMA Dataset

The US Department of Transportation has created a pioneering platooning system that incorporates not only the well-known ACC but also includes V2V and V2I communication. This Cooperative Adaptive Cruise Control (CACC)

[1]https://sites.google.com/view/accresources/home

implementation is founded on the Cooperative Automation Research and Mobility Applications (CARMA) platform. For more information, we refer the reader to the corresponding publications, Tiernan et al. (2017) and Tiernan et al. (2019). CARMA platform facilitates cooperative tactical manoeuvers of automated driving systems with other vehicles and roadway infrastructure through wireless communication. The CARMA platform consists of several components that communicate with each other using the Robot Operating System (ROS) framework. The platform also enables V2V and V2I communications using various communication protocols, such as cellular or dedicated short-range communications (DSRC) messaging. Testing for CARMA2 was also conducted at the US Army's Aberdeen Test Center and involved a fleet of five 2013 Cadillac SRX passenger vehicles. The results in this chapter refer to observations on a 7.2-km two-lane road of the proving ground. Spacing measurements were obtained with PinPoint GPS, while for speed measurements the production wheel speed was used. More details on the dataset can be found in the above-mentioned reports.

Human-Driven Car-Platoon Datasets

Car platoons with human-driven vehicles are also available in the literature. Such datasets are particularly important in order to identify the behavioural differences between human drivers and controllers and provide quantification for the impact of each type of driver. In 2014, a study was carried out in Atlanta, United States, involving a campaign that employed six vehicles on a public two-lane urban street during non-peak hours (Laval et al., 2014). Between 2002 and 2003, a series of vehicle platooning trials, with human drivers at the helm, took place in the vicinity of Naples, Italy (Punzo & Simonelli, 2005). The Beijing Jiaotong University platooning experiment with 25 human-driven vehicles was carried out in 2013 on a 3.2-km stretch of Chuangxin Avenue in a suburban area in Hefei City, China (Jiang et al., 2015). There is no interference from other vehicles. High-precision differential GPS devices were used for each vehicle's speed and position monitoring.

SINGLE ACC DYNAMICS

ACC controllers show different driving behaviours than human drivers and the challenge is to identify these differences and manage to quantify their impact at different levels (vehicle, car-platoon, or traffic). Fig. 2.1 visually shows the behavioural characteristics of ACC and human-driven vehicles. Fig. 2.1(a) shows a platoon of ACC-driven vehicles while Fig. 2.1(b) shows the same platoon but with human-driven vehicles. The driving cycles are very similar and therefore the comparison is feasible. Each figure is split into three rows. The first row is the oblique trajectories for the vehicles in the platoon. Since the intra-platoon space variations are not visible in such a long path, we assume a moving observer. The observer follows the platoon with an average (for this path) speed, which in this cycle's case is 19 m/s. Then all vehicle trajectories are projected on the trajectory of this moving observer. More information about oblique representations can be found in the literature (Cassidy & Windover, 1995; Munoz & Daganzo, 2002). It is clear from the trajectory profiles that different human drivers keep different distances from their leading vehicle and their oscillations around their target spacing (if we assume that there is any, for human drivers) are more prominent. ACC systems demonstrate smoother driving behaviour under stable car-following conditions with short oscillations around their target time gap and homogenous spacing policy. Overall, ACC systems reach smaller headways during perturbation than humans (roughly 0.7 vs 0.9 second). The above observations are validated in the speed profiles where ACC vehicles demonstrate much lower speed variation around the platoon leader's speed. Additionally, the delay from the reaction of ACC systems shows higher speed overshoot compared to human drivers, who produce smoother trajectories during perturbations. Furthermore, there is an effect on the order in the platoon since the last vehicle has the strongest speed dynamics in ACC mode, which eventually leads to higher energy consumption. This phenomenon is discussed in the following paragraphs as platoon string instability. The energy impact is also investigated in a separate section. Finally, the heterogeneity in spacing policy is clear for the human drivers in Fig. 2.1(b), while ACC systems aim for very similar headway even for vehicles of different manufacturers.

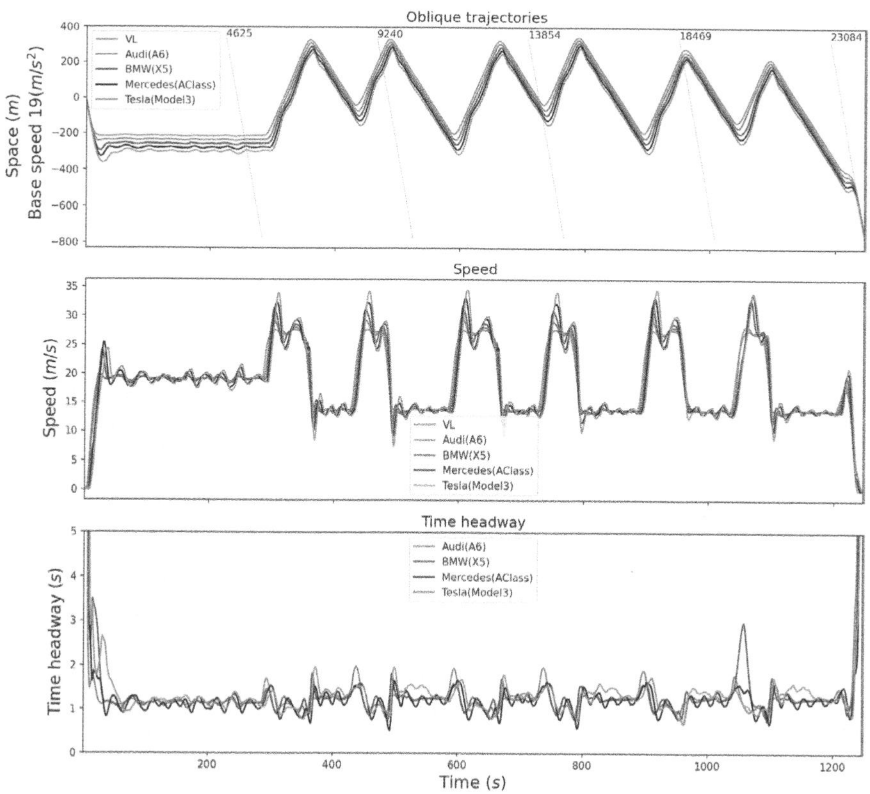

(a)

Fig. 2.1. Visualising the Following Behaviours of Commercial ACC Systems and Human Drivers – Astazero Campaign.

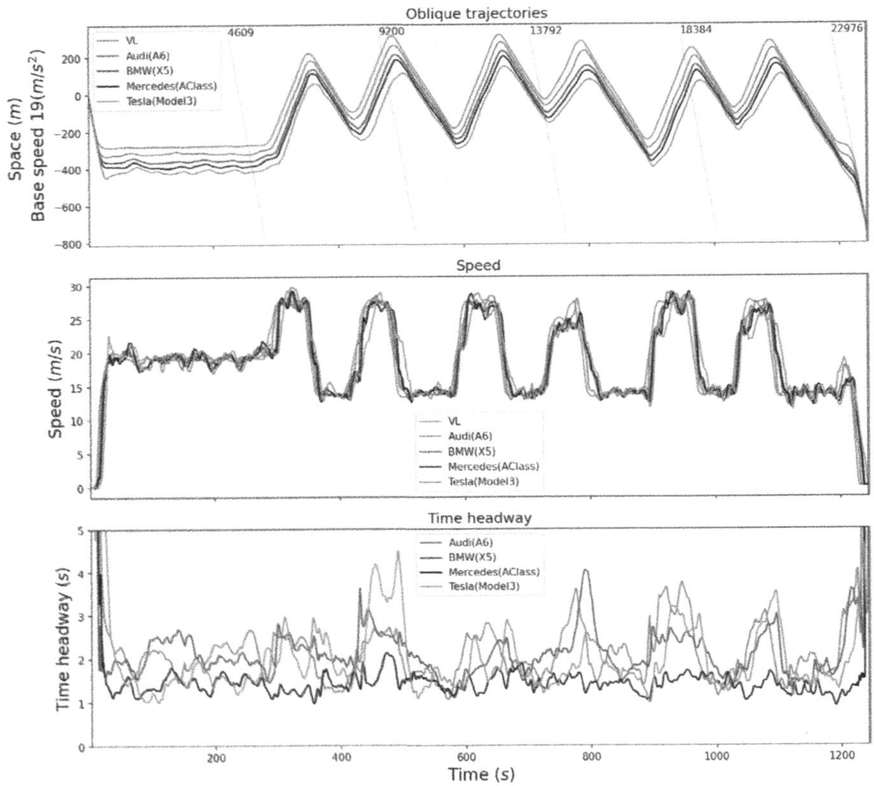

(b)

Fig. 2.1. (Continued)

Furthermore, it is interesting to observe the acceleration dynamics for ACC controllers under different desired settings and compare them with human drivers. Fig. 2.2 shows the acceleration distributions for the ZalaZone campaign included in the OpenACC dataset. The data refer to many experiments performed using ACC-driven vehicles with the minimum desired headway setting, ACC-driven vehicles with the maximum desired headway setting, and human-driven vehicles. Fig. 2.2(a) shows the speed distributions derived from the observations. Not all modes (ACC-min/max and Human) have the same number of observations. However, due to similar experimental design, that is similar driving cycle, all three demonstrate two prevalent speed peaks around 11 and 16 m/s. Nevertheless, please note that the acceleration dynamics might differ for the same driving cycle with alternate modes due to driving behaviour but also other factors such as different reaction times, controller design, and string instability. As we will discuss later in the chapter, string instability for ACC systems under the minimum setting produces higher speed variation after perturbations (over/undershooting) directly leading to increased energy demand. ACC systems under

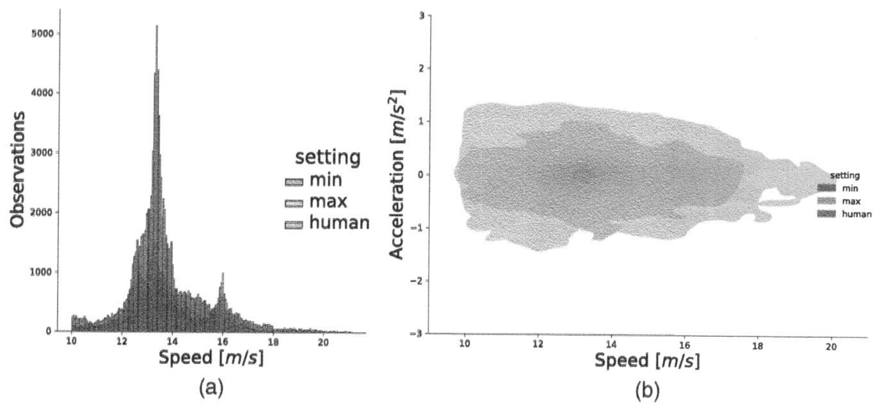

Fig. 2.2. Acceleration Dynamics on the Same Campaign (Multiple Driving Cycles) for ACC-Driven Vehicles with Minimum and Maximum Settings and Human Drivers – ZalaZone Campaign.

the maximum headway setting show similar behaviours to human drivers. The acceleration dynamics change for different speeds due to (among others) variable power distribution capacity of the vehicle over speed. Therefore, it is common to observe lower acceleration values at high speeds than at low ones. Consequently, the comparison of acceleration dynamics at different speeds can be unfair. We refer the reader to the literature for more details on vehicle dynamics and their impact on the observed acceleration (Fadhloun et al., 2015; Makridis et al., 2019; Suarez et al., 2022).

The speed distributions here show that the three driving models (minimum/maximum setting and human driver) have similar desired speeds (by observing the peaks), enabling a fair comparison. The speed distributions are different since the behaviour of each mode is different under similar stimuli. Fig. 2.2(b) illustrates the multivariate kernel density estimation per driving mode. As expected, the higher density is around zero acceleration, corresponding to stable following conditions. However, here we are mostly interested in the speed over acceleration areas. It is clear that ACC with minimum setting led to more extreme acceleration values than the other two driving modes. Therefore, we have serious indications that the ACC system under normal (safety non-critical) conditions can produce much higher and lower accelerations than human drivers. With simple investigations within the data, we can also verify that these findings are independent of the vehicle brand or model, the driver, and the vehicle order. As we will see later in this chapter, such behaviours are related to the string instability of commercial ACC systems.

Headway insights during unstable driving conditions enhance our understanding of how automated systems behave. Fig. 2.3(a) and (b) depicts the speed and headway profiles of vehicles in a platoon, showcasing scenarios involving human-driven, ACC, and CACC systems. Each colour corresponds to a vehicle's profile, where vL (in the caption) signifies the platoon leader, while v1–v4 each of the followers respectively. Of course, the results are dependent on the vehicle model/brand

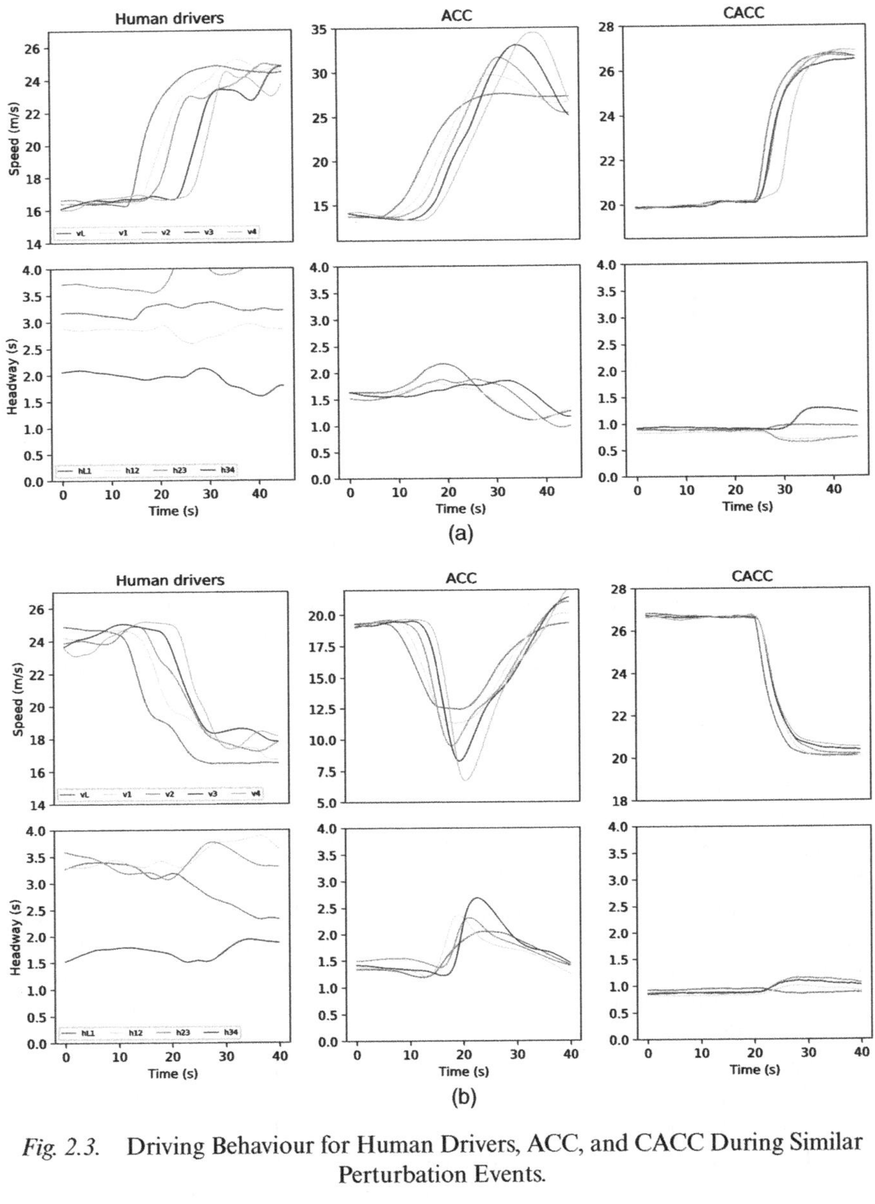

Fig. 2.3. Driving Behaviour for Human Drivers, ACC, and CACC During Similar Perturbation Events.

and the traffic conditions, so these figures can be only considered exemplary ones but also representative of different experiments in the literature. These slices have been intentionally selected to investigate headway patterns during perturbations. Notably, in unstable car-following conditions, we observe that the headway of ACC-equipped vehicles deviates from optimal (target) values due to the overreaction of the controller. During the CARMA2 experiment, there is no overreaction presumably due to the additional information arriving through communication. Finally, human drivers seem to maintain a distance that is not correlated with a specific desired headway. Additionally, human drivers are able to anticipate perturbations downstream and thus demonstrate more stable speed patterns than ACC systems. For the same reason, long platoons of ACC-driven vehicles may lead to huge variations around the desired speed for the last vehicles when a perturbation reaches the platoon. CACC systems seem to achieve the best results from both worlds, that is smaller headways, low-speed variation for all platoon members under stable conditions, the anticipation of perturbations downstream, and low reaction time.

OBSERVABLE RESPONSE TIME

We consider as response or reaction time, the time delay, under stable conditions, between the moment that an action of the vehicle in front occurs to the moment in which the following vehicle reacts. In traffic flow theory, the reaction time is related to the traffic capacity at motorways (Anon, 2011; Calvert et al., 2018). It is challenging to estimate the response time from observations as stable conditions are rarely observed in real-world. Furthermore, the controller design is unknown and the observations contain noise. For real-world cases, the observable response time reflects the time delay experienced by a vehicle within a platoon when reacting to speed changes initiated by a leading vehicle, as witnessed through empirical data measurements. Several techniques estimate the observable response time. Inferred values might vary even within the same experiment since methods have to deal with data noise and the fact that the operation principles of commercial ACC are not publicly available. Therefore, a comparison between the accuracy of the methods used here is out of this chapter's scope. For more information on this, we refer the reader to the corresponding paper (Brunner et al., 2022). For human drivers, it is more challenging to infer the response time (reaction time) since they do not follow a constant headway policy like ACC systems. This means that even when we observe a stimulus on the trajectory of the leader, we cannot be certain that the delay in the response of the follower is the reaction time, that is it can also include delays due to the driving behaviour of the follower. Therefore, this chapter does not provide an analysis of the reaction time of human drivers.

Existing literature indicates that the average human reaction time is approximately 1.2 seconds (Lanaud et al., 2021). This value also holds true for ACC systems (Li et al., 2021; Makridis et al., 2019). For CACC systems response time estimates are scarce in the literature (Brunner et al., 2022). Estimating the response time on worldwide observations reveals very interesting findings. More specifically, three different state-of-the-art techniques in the literature were used (Lanaud et al., 2021; Li et al., 2021; Makridis et al., 2019). The objective is to gain insights independent of the acquisition technology, the driving cycle particularities and the efficiency of the computational approach.

The distributions of estimated response times per dataset for ACC and CACC vehicles are shown in Fig. 2.4. The response times are grouped based on the headway setting. The setting information is shown in the *x*-axis labels, wherever necessary. In the AstaZero campaign, all ACC-driven vehicles use the minimum

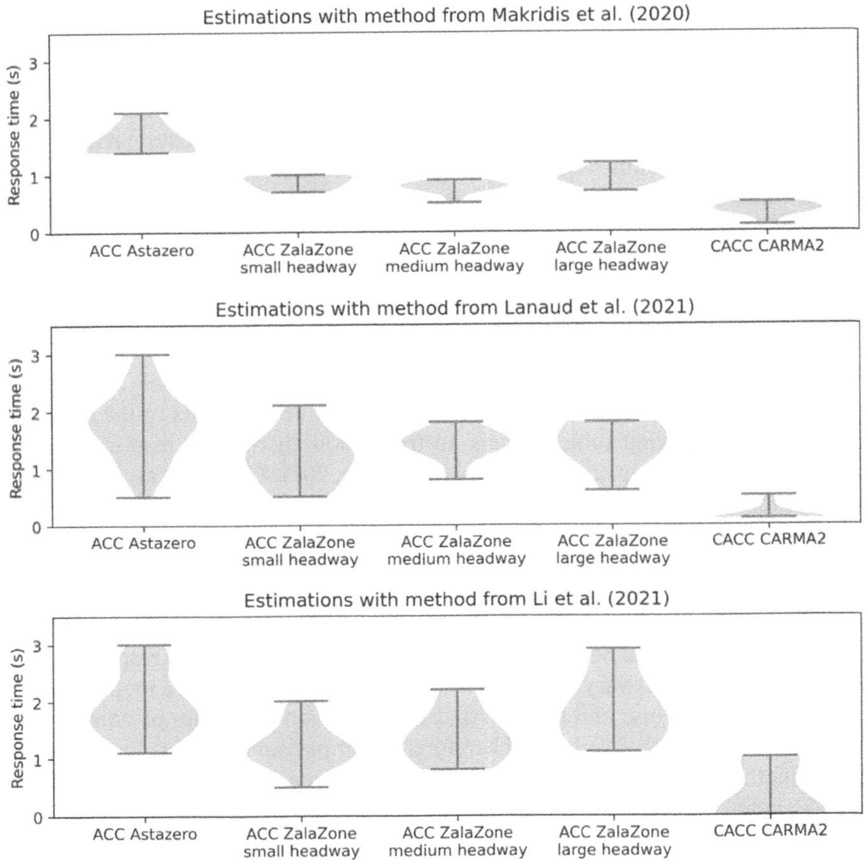

Fig. 2.4. Estimated Observable Response Times for ACC- and CACC-Driven Vehicles Based on Worldwide Observations.

ACC setting. In the ZalaZone campaign, there are observations with all three settings (see labels), and in the CARMA campaign, there is only a single setting for CACC vehicles. The response time is assessed on 35 partial trajectories that correspond to isolated perturbation events, i.e. stable car-following conditions followed by a sudden increase or decrease in the leader's speed. The estimated response times for the ACC vehicles range from approximately 1.3 to 2.5 seconds. Notably, the CARMA2 CACC platoon showcases significantly shorter response times, with values falling between 0 and 1 second. We can derive only preliminary indications from this comparison as CACC data are few in the literature and the technology is not yet market-ready. Furthermore, connectivity inherits a set of potential safety risks that are not discussed in this chapter. Nevertheless, distributing dynamically the vehicle-specific speed and position information to other road users through vehicle connectivity shows the potential to reduce response times and increase link capacity without compromising safety.

STRING STABILITY

Platoon string stability is related to both safety and traffic efficiency. As outlined in the study by Monteil et al. (2019), string stability in a platoon is established when the last following vehicle successfully mitigates the perturbation introduced by the lead vehicle, called weak string stability. Gunter, Gloudemans, et al. (2020) highlight that unstable behaviour within vehicle platoons can potentially trigger stop-and-go traffic patterns and the emergence of phantom traffic jams. Studies in the literature indicate that ACC-equipped vehicles adhering to a constant time-gap policy can attain string stability. However, observation-based studies (Ciuffo et al., 2021; Gunter, Gloudemans, et al., 2020; Makridis et al., 2020), have concluded that ACC platoons in commercial vehicles demonstrate (at least) weak string instability.

The stability behaviour of CACC-driven vehicles is only slightly investigated through experiments (Brunner et al., 2022) due to very few available observations. A number of perturbation events, that is stable car-following followed by a change in the state of the leading vehicle, were classified into two classes, acceleration and deceleration perturbations, depending on whether stable conditions are followed by an acceleration or deceleration of the leader. In the left subplots of Fig. 2.5(a) and (b), we observe the equilibrium speeds and the perturbation magnitudes. The right subplots of Fig. 2.5(a) and (b) illustrate the magnitude

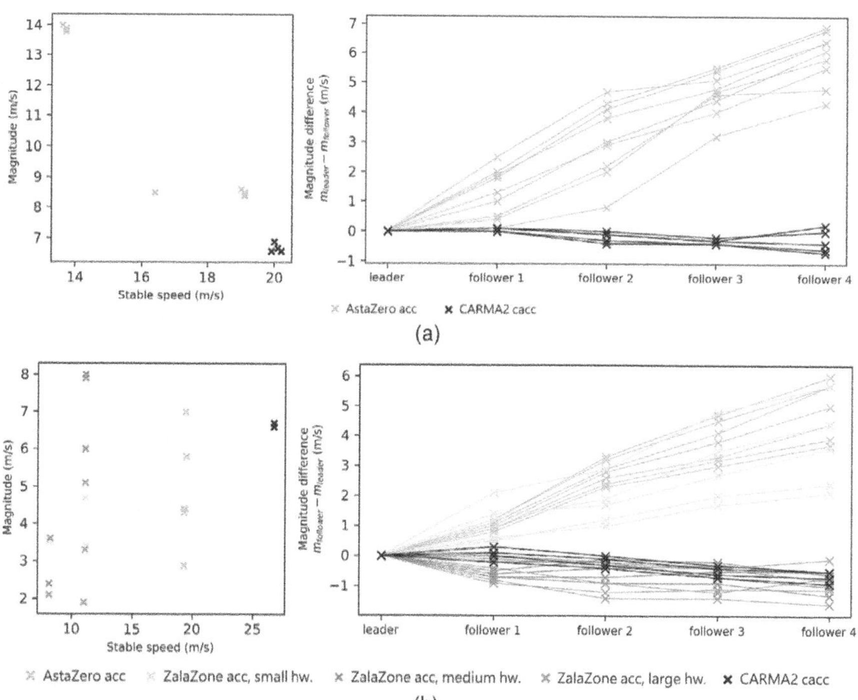

Fig. 2.5. Stability Representation During Acceleration (a) and Deceleration (b) Perturbation Events.

differences between the leading vehicle and the following vehicles. If this ratio amplifies it corresponds to string instability. Conversely, if the followers' magnitudes are below zero, the perturbation dissipates, and the platoon remains stable. Consequently, it's observed that ACC platoons with tighter headway settings tend to exhibit string instability, whereas ACC platoons with medium or large headway settings, as well as the CACC platoon, maintain stability.

TRAFFIC FLOW

The increasing number of observations provides unique opportunities to study the traffic flow characteristics that arise from this technology. However, it should be noted that the nominal capacity based on platoon observations might differ a lot from the actual capacity observed on public roads. This has to do with two main factors, the platoon length that is usually small and the fact that in real-world conditions there are significant perturbations due to on-ramp inflows, as well as discretionary and mandatory lane changing. A typical capacity for human drivers on highway conditions is around 1,500–1,800 veh/h and the nominal capacity of ACC-driven vehicles varies depending on the ACC setting (Chiabaut et al., 2009; Shi & Li, 2021). For the minimum ACC setting, the nominal capacity of ACC platoons reaches almost double the previous value, i.e. around 3,000 veh/h. For ACC long setting, the nominal capacity is in the same range as for human drivers (Li et al., 2022; Makridis & Kouvelas, 2022; Shi & Li, 2021). A very recent study on this topic (Li et al., 2022) provided experimental evidence to this knowledge base and presented a comprehensive empirical study on ACC equilibrium behaviours via the resulting fundamental diagrams. It is shown that, like human-driven vehicles, ACC systems display a linear equilibrium spacing-speed relationship (within the range of available data) but the key parameters of these relationships can differ significantly from human-driven traffic depending on input settings. At the minimum ACC headway setting, equilibrium capacities above 3,500 vehicles per hour are observed, together with an extremely fast equilibrium wave speed of 100kms/hour. These fast waves are unfamiliar to human drivers and may pose a safety risk. The results also suggest that ACC jam spacing can be much larger than in human traffic, which reduces the network storage capacity. Investigations on CACC technology are not yet available due to few available experimental observations, however, connectivity among the vehicles is expected to shift critical density to higher values and increase link capacity in comparison to human drivers and ACC systems.

TRAFFIC HYSTERESIS

Traffic oscillations emerge in road traffic as a result of lane changes, abrupt driver manoeuvers (Ahn & Cassidy, 2007), or even the string unstable behaviour of human drivers in dense traffic conditions (Sugiyama et al., 2008). During those oscillations, it has been observed that often the flow during the acceleration is smaller than the flow during the deceleration part of the oscillation

(Yeo & Skabardonis, 2009). On a density/flow or speed/flow diagram, this shows as a hysteresis loop. Hysteretic recovery of the steady-state speed decreases the throughput after an oscillation, with negative impacts on the traffic conditions. The most appropriate way to investigate hysteresis from platoon trajectories is to define a space-time area around the platoon that estimates stationary conditions (Laval, 2011). Stationary conditions include isolated traffic states and consequently, the aggregation of different detected states can help form the fundamental diagram. As noted in the above paper, both positive and negative hysteresis loops can be observed. Furthermore, recent studies investigate the relationship between hysteresis and string instability observed in platoon experiments (Albano et al., 2021). Finally, it should be noted that traffic hysteresis is a collective traffic phenomenon that is closely related to driving dynamics and driver heterogeneity, therefore, without a theoretical optimum for traffic flow.

In this work, we investigate the emergence of hysteresis loops in platoons of ACC or CACC vehicles that encounter a perturbation. For the ACC vehicles, two perturbations have been chosen from the AstaZero dataset, one with vehicles using the smallest time-gap setting and one with vehicles using the longest time-gap setting. The CACC data are taken from one of the perturbations of the CARMA dataset. The flow, density, and speed can be calculated using Edie's generalised definitions (Edie, 1961), as proposed in the work of Laval (2011):

$$k=\sum_{i=0}^{n} t_i / |A|$$

$$q=\sum_{i=0}^{n} x_i / |A|$$

$$v=\frac{q}{k}=\sum_{i=0}^{n} x_i / \sum_{i=0}^{n} t_i$$

where the symbols k, q and v stand for density, flow, and speed in region A, respectively, $|A|$ is the area of A, and t_i, x_i are the ith vehicle travel time and distance travelled inside A, respectively. Region A is chosen in a way that maximises the chances of having stationary conditions for the platoon. More specifically, the boundaries of the region A correspond to the wave speed w, retaining consistency with kinematic wave theory.

The speed/flow diagrams are presented in Fig. 2.6(a) for the ACC with the shortest time-gap setting, Fig. 2.6(b) for the ACC with the longest time-gap setting, and Fig. 2.7 for the CACC platoons. The platoons start from a velocity around 90 km/h and decelerate to a slower speed, with the deceleration being imposed by the leading vehicle. The leading vehicle maintains this low speed for a time and then accelerates back to the initial speed. The drop in the CACC case is smaller than the drop in the ACC cases, and this is expected to affect the magnitude of the hysteresis measured, but the qualitative results are not expected to change significantly. The deceleration branch is denoted by the continuous red line, blue dotted lines are used for the acceleration part, and the dashed green line denotes the part of stabilising around a low-speed steady state condition.

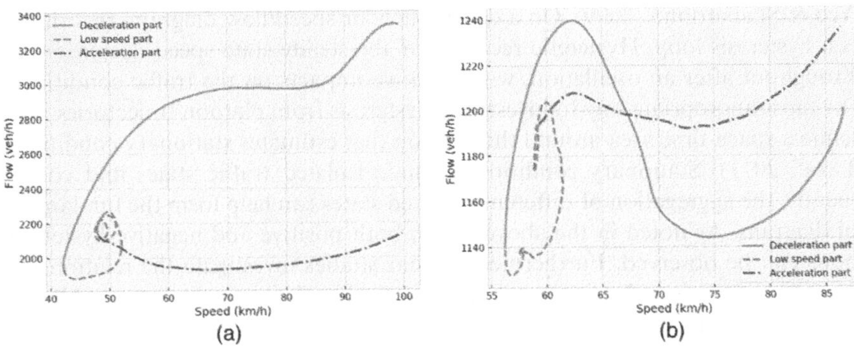

Fig. 2.6. Speed/Flow Diagrams for Perturbations of a Platoon of ACC Vehicles Using the (a) Shortest Time-Gap, and (b) Longest Time-Gap Setting.

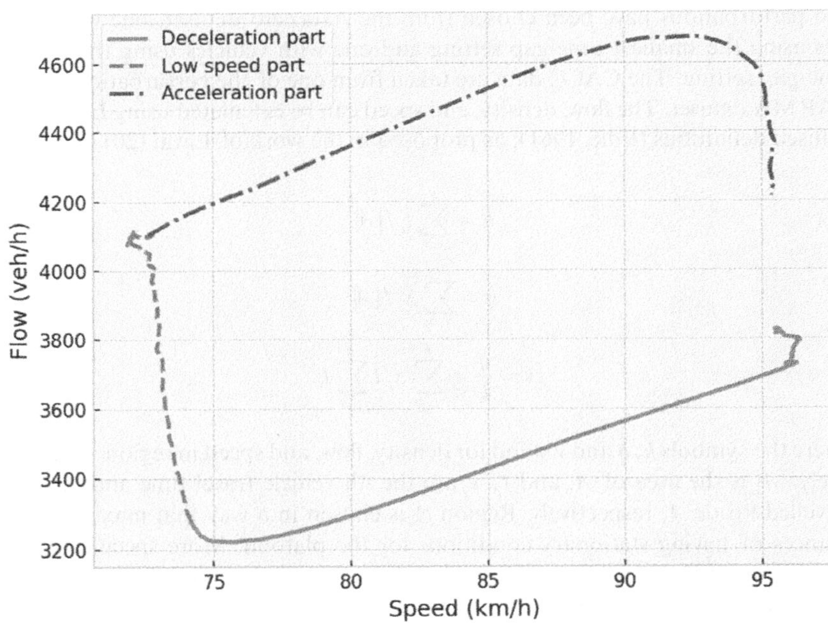

Fig. 2.7. Speed/flow Diagrams for Perturbations of a Platoon of CACC Vehicles.

In Fig. 2.6(a), the initial flow is more than 3,000 veh/h, which is higher than the flow human-driven vehicles are shown to produce. However, there is a steep drop during the deceleration of the platoon. The platoon is slow to stabilise in the lower speed value, which is shown by the green circles in the graph. In the acceleration branch, as the speed increases, the flow remains at lower values. Longer time-gap settings have been shown to produce less unstable platoons or even string-stable ones. This stability seems to have an effect on the hysteresis loop, as shown in Fig. 2.6(b), and there is no obvious degradation of the flow during the acceleration phase. However, the flow is very limited, even for steady-state conditions. On the contrary, for the CACC platoons that use cooperation through connectivity to improve the platoon dynamics, the flow is in very high values. While there is a drop during the deceleration phase, the acceleration branch is much larger, reaching values over 4,000 veh/h. The system takes advantage of the small inter-vehicle gaps, to recover the speed fast and efficiently, increasing the throughput after a potential bottleneck. Additionally, the examples in Fig. 2.6 correspond to positive and negative hysteresis loops as pointed out in the literature, see Laval (2011), something also observed after the analysis of various heterogeneous datasets (Makridis & Kouvelas, 2022). The appearance is related less to the driving mode and more to the intra-platoon driving behaviours. Finally, it seems like in the short-time gap case the traffic jam builds quickly while its resolution is much slower. In the long-time gap case, the resolution of the traffic jam is much faster, and normal (even better than initial) conditions are restored. Observations in the literature show that there is an interplay with counter-balancing effects and no clear winner for different ACC headway settings, between reaction time, string instability, and traffic hysteresis (Albano et al., 2021; Brunner et al., 2022).

ENERGY CONSUMPTION

There is evidence in the literature that ACC-driven vehicles are more energy efficient than human drivers under stable car-following conditions (lower speed variation) but less energy efficient under unstable car-following conditions (string instability) (He et al., 2020).

To offer quantified results, He et al. (2020) utilise tractive energy consumption as an indicator of the energy impact. Tractive energy does not consider the efficiency of the powertrain (electric, internal combustion engine or other). Employing the same methodology for ACC and CACC systems ensures comparability, and the study focuses on segments of experimental data where the platoon's speed profile closely aligns across the different modes.

The energy consumption, which is the sum of tractive power over the distance covered, is calculated for various scenarios and presented in Fig. 2.8. The left and middle subplots reveal that ACC platoons, on average, consume more energy than human drivers. In the right subplot, an ACC platoon is contrasted with two CACC platoons. While the vehicle platoon from the CARMA1 campaign does not offer improvements in terms of energy consumption, the CACC platoon from the CARMA2 campaign demonstrates, on average, lower energy usage compared to the ACC platoon.

The coloured bullets in Fig. 2.8 represent the position of the vehicles in the car-platoon, starting with the blue bullet and ending with the purple bullet. The response time results are highly correlated with the driving mode and the corresponding headway settings. The first two columns refer to the AstaZero campaign which refers to ACC-driven platoons with the minimum desired setting only. Energy consumption increases as we look upstream of the ACC platoon. The opposite holds for CACC platoons. This is probably related to unstable versus

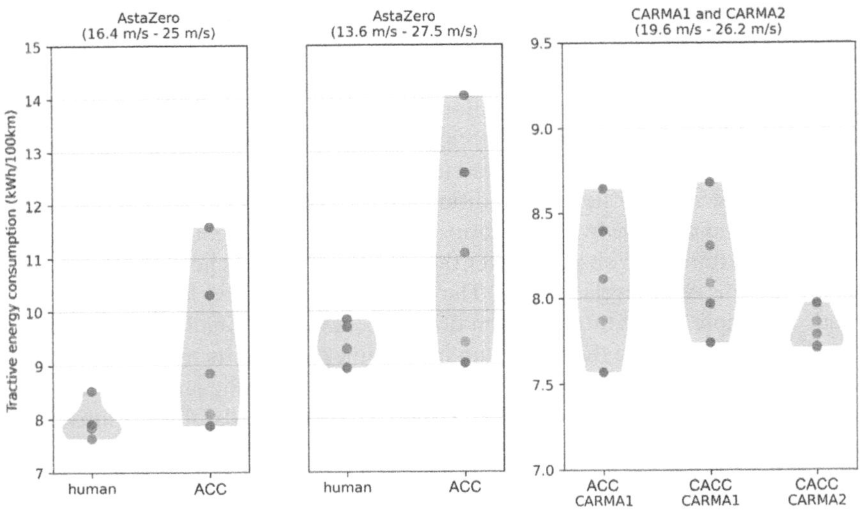

Fig. 2.8. Energy Consumption Comparison of the Platoons in Three Different Scenarios.

stable platoon behaviour. Furthermore, discussing the energy demand per vehicle we can see that the behaviour of an individual ACC-driven vehicle is less interesting than the behaviour of a car-platoon with ACC-driven vehicles. String instability of the platoon increases a lot the energy demand as we observe upstream on the platoon.

TRAFFIC SAFETY

All observations refer to vehicles of automation level 2. As such, there is a necessity for a driver to be in place to monitor the environment and take back control in case of an emergency. Therefore, the emergency reactions of existing ACC systems are of little to no consequence, as they are not expected to avoid a collision. In most cases, the deceleration or steering actions are bounded within a set of comfortable values, so they cannot initiate sharp reactions that may disturb the passengers, even if it is necessary. In the last case, different systems such as Automatic Emergency Braking (AEB) might take over, with the responsibility still staying with the human driver. However, this does not completely exempt such systems from affecting traffic safety. It is now becoming apparent that careful human drivers can take pre-emptive actions to avoid emergencies before harsh reactions are necessary. The term “Tactical safety” has been coined to describe such decisions that are made on a tactical level to retain safety (Schoener, 2020). On the contrary, improper vehicle behaviour can create unsafe conditions or increase the severity of existing ones, before reaching the point when a reaction is necessary that is the point after which a level 2 system may be ill-equipped to react.

Data from the AstaZero campaign have been used to evaluate this characteristic using a Proactive Fuzzy surrogate Safety metric (PFS), finding that the ACC systems, when using the minimum desired time gap would sometimes enter situations that are proactively unsafe (Mattas et al., 2020). In those situations, a harsh deceleration from the leader vehicle would lead to an accident, as there would not be enough space for a safe reaction. Those cases do not necessarily have to do with the steady-state inter-vehicle distance but may come up when a platoon is in a transient state. Sting unstable behaviour is probably the most important example of such improper driving behaviour. Using the data from the ZalaZone campaign, the Time To Collision (TTC) metric is used to identify conflicts. The results from the Dynamic platform are presented in Fig. 2.9(a) for the shortest time-gap and Fig. 2.9(b) for the longest time-gap setting, in the form of cumulative distribution plots. The lines are coloured according to the position of the vehicle in the platoon, with green colours indicating the vehicles in the front of the platoon and red colours indicating vehicles in the tail of the platoon. For the shortest time-gap platoons, the last vehicles show a large frequency of TTC values measured that are less than 1.5 seconds and denote severe conflicts. It is important to note that the initial perturbation of the leading vehicle was mild, and the situation becomes dire as the perturbation magnitude increases while travelling upstream. On the other hand, for the long time gap, the last vehicles never experienced TTC values as small, and the distribution shows larger TTC values than the first vehicles in the platoon.

Looking forward to the large-scale introduction of vehicles with higher levels of automation into the market, it has to be clear that safety does not have to do only with the reaction to emergencies, and the driving behaviour in normal traffic conditions can lead to the emergence or decrease of hazardous traffic conflicts. Focusing on normal traffic behaviour, the need for string stability is thus becoming imperative, and therefore it can be one of the requirements in new regulations on highly automated systems. More research is needed to identify methods to validate the string stable behaviour of a commercial automated driving system by track testing, or real-world tests.

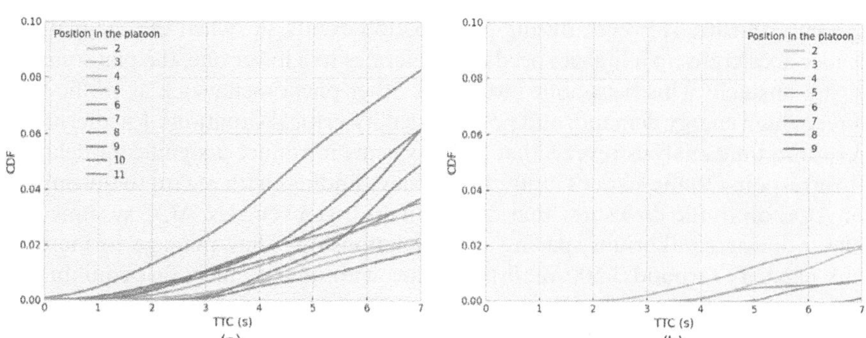

Fig. 2.9. Cumulative Distribution of TTC Values During Perturbations for Platoons Using (a) the Shortest Time-Gap and (b) Longest Time-Gap.

DISCUSSION AND CONCLUSION

The objective of this chapter has been to enhance our comprehension of advanced automated systems like ACC and CACC through empirical observations. Our aim was to compare the different controllers, identify commonalities and differences with human drivers and understand potential impacts on the safety and efficiency of road transport networks. Car-platoon data from empirical observations of independent vehicle platooning experiments have been used to provide quantitative results on systems' performance. Worldwide datasets with observations are discussed and analysed towards this goal. The analysis spans from individual system-equipped vehicles to car-platoon behaviours with regard to dynamics, system response time, string stability, traffic hysteresis, safety, and energy efficiency.

With new regulations regarding the approval of automated driving systems being now adopted or drafted, an effort is made to guarantee the best of the potential outcomes regarding the effects of those systems on traffic. Their operation is controlled and restricted by regulation requirements, to ensure safe driving behaviour. For the establishment of those requirements, it is imperative that the physical capabilities of such systems are well understood. Moreover, such restrictions could have adverse implications on traffic conditions. Thus, a systematical and empirical investigation of the driving characteristics of the existing automated systems, as presented in the current work, is necessary and must be carried out continuously. The results of those investigations are vital for research, development, and regulation to be informed and aware of the conditions and to drive towards upgrading the state of traffic networks. Automated Driving Systems, and smart mobility solutions in general, are considered among the tools to achieve not only a safer but also a more efficient and greener transport system. They are among the technologies asked to contribute to the so-called "Twin Transition" towards a green and digital society (European Commission Joint Research Centre, 2022). A reality check about whether such technologies are really contributing to both transitions is therefore very relevant at present.

ACC systems generate less speed and headway variations under stable following conditions in comparison to human drivers, even for large car platoons of such vehicles. However, during perturbation events, i.e. when the car-platoon leader accelerates to a higher speed or decelerates to a lower one, the platoons are string unstable. This behaviour introduces other phenomena such as traffic hysteresis, high energy demand, and potential safety-critical situations. Furthermore, response time analysis reveals that ACC systems introduce a significant delay in their response to the leader's input. Preliminary findings with regard to the impact of ACC on traffic flow show that, like human-driven vehicles, ACC systems display a linear equilibrium spacing-speed relationship. They manage to increase link capacity (around 3,500 veh/h), together with an extremely fast equilibrium wave speed of 100 kms/h. Connectivity in ACC systems (CACC) mitigates most downsides of ACC leading to string stability, shorter inter-vehicle spacings, smoother speed profiles, and less energy demand, without compromising traffic safety. Therefore, CACC technologies hold stronger potential to contribute to more sustainable transport systems, although the energy requirements to move from ACC to widely adopted CACC remain under-studied and is an interesting issue for future research.

REFERENCES

Ahn, S., & Cassidy, M. J. (2007). Freeway traffic oscillations and vehicle lane-change maneuvers. *Transportation and Traffic Theory*, *1*, 691–710.

Albano, G., Mattas, K., & Ciuffo, B. (2021). *The relationship between hysteresis and string stability in traffic flow. A simulation-based investigation.*

Anon. (2011). *75 Years of the fundamental diagram for traffic flow theory: Greenshields Symposium: July 8-10, 2008, Woods Hole, Massachusetts.* Transportation Research Board.

Brunner, J. S., Makridis, M. A., & Kouvelas, A. (2022). Comparing the observable response times of ACC and CACC systems. *IEEE Transactions on Intelligent Transportation Systems*, 1–10. https://doi.org/10.1109/TITS.2022.3165648

Calvert, S. C., Femke, L., van Wageningen-Kessels, M., & Hoogendoorn, S. P. (2018). Capacity drop through reaction times in heterogeneous traffic. *Journal of Traffic and Transportation Engineering (English Edition)*, *5*(2), 96–104. https://doi.org/10.1016/j.jtte.2017.07.008

Cassidy, M. J., & Windover, J. R. (1995). Methodology for assessing dynamics of freeway traffic flow. *Transportation Research Record*, *1484*, 73–79.

Chiabaut, N., Buisson, C., & Leclercq, L. (2009). Fundamental diagram estimation through passing rate measurements in congestion. *IEEE Transactions on Intelligent Transportation Systems*, *10*(2), 355–359. https://doi.org/10.1109/TITS.2009.2018963

Ciuffo, B., Mattas, K., Makridis, M., Albano, G., Anesiadou, A., He, Y., Josvai, S., Komnos, D., Pataki, M., Vass, S., & Szalay, Z. (2021). Requiem on the positive effects of commercial adaptive cruise control on motorway traffic and recommendations for future automated driving systems. *Transportation Research Part C: Emerging Technologies*, *130*, 103305. https://doi.org/10.1016/j.trc.2021.103305

EC. (2022). Commission Implementing Regulation (EU) 2022/1426 of 5 August 2022 Laying down Rules for the Application of Regulation (EU) 2019/2144 of the European Parliament and of the Council as Regards Uniform Procedures and Technical Specifications for the Type-Approval of the Automated Driving System (ADS) of *Fully Automated Vehicles (Text with EEA Relevance)*, Vol. 221.

Edie, L. C. (1961). Car-following and steady-state theory for noncongested traffic. *Operations Research*, *9*(1), 66–76. https://doi.org/10.1287/opre.9.1.66

European Commission. Joint Research Centre. (2022). *Towards a green & digital future: Key requirements for successful twin transitions in the European Union.* Publications Office.

Fadhloun, K., Rakha, H., Loulizi, A., & Abdelkefi, A. (2015). Vehicle dynamics model for estimating typical vehicle accelerations. *Transportation Research Record*, *2491*(1), 61–71. https://doi.org/10.3141/2491-07

Gunter, G., Gloudemans, D., Stern, R. E., McQuade, S., Bhadani, R., Bunting, M., Delle Monache, M. L., Lysecky, R., Seibold, B., Sprinkle, J., Piccoli, B., & Work, D. B. (2020). Are commercially implemented adaptive cruise control systems string stable? *IEEE Transactions on Intelligent Transportation Systems*, 1–12. https://doi.org/10.1109/TITS.2020.3000682

Gunter, G., Janssen, C., Barbour, W., Stern, R. E., & Work, D. B. (2020). Model-based string stability of adaptive cruise control systems using field data. *IEEE Transactions on Intelligent Vehicles*, *5*(1), 90–99. https://doi.org/10.1109/TIV.2019.2955368

Hans-Peter, S. (2020). *Challenging highway scenarios beyond collision avoidance for autonomous vehicle certification.* https://doi.org/10.13140/RG.2.2.29355.05926

He, Y., Ciuffo, B., Zhou, Q., Makridis, M., Mattas, K., Li, J., Li, Z., Yan, F., & Xu, H. (2019). Adaptive cruise control strategies implemented on experimental vehicles: A review. *IFAC-PapersOnLine*, *52*(5), 21–27. https://doi.org/10.1016/j.ifacol.2019.09.004

He, Y., Makridis, M., Fontaras, G., Mattas, K., Xu, H., & Ciuffo, B. (2020). The energy impact of adaptive cruise control in real-world highway multiple-car-following scenarios. *European Transport Research Review*, *12*(1), 17. https://doi.org/10.1186/s12544-020-00406-w

Jiang, R., Hu, M.-B., Zhang, H. M., Gao, Z.-Y., Jia, B., & Wu, Q-S. (2015). On some experimental features of car-following behavior and how to model them. *Transportation Research Part B: Methodological*, *80*, 338–354. https://doi.org/10.1016/j.trb.2015.08.003

JRC-EC. (2020). *OpenACC Database*. https://data.jrc.ec.europa.eu/dataset/9702c950-c80f-4d2f-982f-44d06ea0009f

Lanaud, E., Ladino, A., & Buisson, C. (2021). First observations about response times and connectivity in a vehicles platooning experiment. *Findings*, 21190. https://doi.org/10.32866/001c.21190

Laval, J. A. (2011). Hysteresis in traffic flow revisited: An improved measurement method. *Transportation Research Part B: Methodological*, *45*(2), 385–391. https://doi.org/10.1016/j.trb.2010.07.006

Laval, J. A., Toth, C. S., & Zhou, Y. (2014). A parsimonious model for the formation of oscillations in car-following models. *Transportation Research Part B: Methodological*, *70*, 228–238. https://doi.org/10.1016/j.trb.2014.09.004

Li, T., Chen, D., Zhou, H., Laval, J., & Xie, Y. (2021). Car-following behavior characteristics of adaptive cruise control vehicles based on empirical experiments. *Transportation Research Part B: Methodological*, *147*, 67–91. https://doi.org/10.1016/j.trb.2021.03.003

Li, T., Chen, D., Zhou, H., Xie, Y., & Laval, J. (2022). Fundamental diagrams of commercial adaptive cruise control: Worldwide experimental evidence. *Transportation Research Part C: Emerging Technologies*, *134*, 103458. https://doi.org/10.1016/j.trc.2021.103458

Makridis, M., Fontaras, G., Ciuffo, B., & Mattas, K. (2019). Mfc free-flow model: Introducing vehicle dynamics in microsimulation. *Transportation Research Record*, *2673*(4), 762–777. https://doi.org/10.1177/0361198119838515

Makridis, M., & Kouvelas, A. (2022). Introducing the platoon fundamental diagram for automated vehicles based on large-scale empirical observations. In *Abstract Book: 10th Symposium of the European Association for Research in Transport (hEART 2022)*. European Association for Research in Transportation.

Makridis, M. A., & Kouvelas, A. (2023). Adaptive physics-informed trajectory reconstruction exploiting driver behavior and car dynamics. *Scientific Reports*, *13*(1), 1121. https://doi.org/10.1038/s41598-023-28202-1

Makridis, M., Mattas, K., Anesiadou, A., & Ciuffo, B. (2021). OpenACC. An open database of car-following experiments to study the properties of commercial ACC systems. *Transportation Research Part C: Emerging Technologies*, *125*, 103047. https://doi.org/10.1016/j.trc.2021.103047

Makridis, M., Mattas, K., & Ciuffo, B. (2019). Response time and time headway of an adaptive cruise control. An empirical characterization and potential impacts on road capacity. *IEEE Transactions on Intelligent Transportation Systems*, 1–10. https://doi.org/10.1109/TITS.2019.2948646

Makridis, M., Mattas, K., Ciuffo, B., Re, F., Kriston, A., Minarini, F., & Rognelund, G. (2020). Empirical study on the properties of adaptive cruise control systems and their impact on traffic flow and string stability. *Transportation Research Record*, 0361198120911047. https://doi.org/10.1177/0361198120911047

Mattas, K., Makridis, M., Botzoris, G., Kriston, A., Minarini, F., Papadopoulos, B., Re, F., Rognelund, G., & Ciuffo, B. (2020). Fuzzy surrogate safety metrics for real-time assessment of rear-end collision risk. A study based on empirical observations. *Accident Analysis & Prevention*, *148*, 105794. https://doi.org/10.1016/j.aap.2020.105794

Monteil, J., Bouroche, M., & Leith, D. J. (2019). L2 and L-infinite stability analysis of heterogeneous traffic with application to parameter optimization for the control of automated vehicles. *IEEE Transactions on Control Systems Technology*, *27*(3), 934–949. https://doi.org/10.1109/TCST.2018.2808909

Munoz, J. C., & Daganzo, C. F. (2002). *Fingerprinting traffic from static freeway sensors*. University of California Transportation Center. https://escholarship.org/uc/item/1mf4n2w8

Punzo, V., & Simonelli, F. (2005). Analysis and comparison of microscopic traffic flow models with real traffic microscopic data. *Transportation Research Record: Journal of the Transportation Research Board*, *1934*, 53–63. https://doi.org/10.3141/1934-06

SAE. (2020). *Taxonomy and definitions for terms related to cooperative driving automation for on-road motor vehicles*. SAE International. https://doi.org/10.4271/J3216_202005

Schoener, H. P. (2020). *Challenging highway scenarios beyond collision avoidance for autonomous vehicle certification*. https://doi.org/10.13140/RG.2.2.29355.05926

Shi, X., & Li, X. (2021). Constructing a fundamental diagram for traffic flow with automated vehicles: Methodology and demonstration. *Transportation Research Part B: Methodological, 150*, 279–292. https://doi.org/10.1016/j.trb.2021.06.011

Suarez, J., Makridis, M., Anesiadou, A., Komnos, D., Ciuffo, B., & Fontaras, G. (2022). Benchmarking the driver acceleration impact on vehicle energy consumption and CO2 emissions. *Transportation Research Part D: Transport and Environment, 107*, 103282. https://doi.org/10.1016/j.trd.2022.103282

Sugiyama, Y., Fukui, M., Kikuchi, M., Hasebe, K., Nakayama, A., Nishinari, K., Tadaki, S-I., & Yukawa, S. (2008). Traffic jams without bottlenecks – Experimental evidence for the physical mechanism of the formation of a jam. *New Journal of Physics, 10*(3), 033001. https://doi.org/10.1088/1367-2630/10/3/033001

Tiernan, T., Bujanovic, P., Azeredo, P., Najm, W. G., & Lochrane, T. (2019). *CARMA testing and evaluation of research mobility applications*. Federal Highway Administration, U.S. DoT. https://doi.org/10.21949/1503647

Tiernan, T. A., Richardson, N., Azeredo, P., Najm, W. G., & Lochrane, T. (2017). *Test and evaluation of vehicle platooning proof-of-concept based on cooperative adaptive cruise control*. Federal Highway Administration, U.S. DoT. https://doi.org/10.21949/1503647

UNECE. (2022). *Proposal for the 01 series of amendments to UN regulation no. 157 (Automated lane keeping systems)*. https://unece.org/sites/default/files/2022-05/ECE-TRANS-WP.29-2022-59r1e.pdf

Yeo, H., & Skabardonis, A. (2009). Understanding stop-and-go traffic in view of asymmetric traffic theory. In W. H. K. Lam, S. C. Wong, & H. K. Lo (Eds.), *Transportation and traffic theory 2009: Golden Jubilee* (pp. 99–115). Springer US.

CHAPTER 3

EVALUATING SUSTAINABILITY IN SIMULATIONS OF AUTOMATED AND CONNECTED TRANSPORT

Nima Dadashzadeh[a], Serio Agriesti[b,c], Hashmatullah Sadid[d], Arnór B. Elvarsson[e], Claudio Roncoli[b] and Constantinos Antoniou[d]

[a]*Logistics, Transport, Operations, and Analytics Group, Huddersfield Business School, The University of Huddersfield, UK*
[b]*Department of Built Environment, School of Engineering, Aalto University, Espoo, Finland*
[c]*FinEst Centre for Smart Cities, Tallinn University of Technology, Tallinn, Estonia*
[d]*TUM School of Engineering and Design, Technical University of Munich, Munich, Germany*
[e]*Infrastructure Management Group, Department of Civil, Environmental and Geomatics Engineering, ETH Zürich, Zürich, Switzerland*

ABSTRACT

Early studies projected potential societal, economic and environmental benefits by the widespread deployment of Autonomous and Connected Transport (ACT) promising a significant reduction of transport costs and improvement in road safety. An effective way of assessing ACT impact is via simulations, where results are largely affected by the scenarios defining the ACT development. However, modelled scenarios are very diverse due to the huge uncertainty in ACT development and deployment. This chapter aims to shed light on the different ACT simulation scenarios and sustainability aspects that should be considered while developing or reporting the simulation results. To this end, this chapter discusses the various simulation approaches, what the required (or the typically utilised)

Sustainable Automated and Connected Transport
Transport and Sustainability, Volume 19, 47–64

ISSN: 2044-9941/doi:10.1108/S2044-994120240000019003

pipelines are, and how some components are more important or less important than in 'classic' modelling and simulation approaches. Special focus is dedicated to the uncertainty related to ACT operational parameters and how these will impact transport modelling. To address said uncertainty, an analysis of current approaches to scenario building is provided, as the chapter guides the reader through different methodologies and clusters them in relation to the desired indicators. Finally, the chapter identifies and proposes Key Performance Indicators (KPIs) that are useful when applying simulation tools to assess ACT scenarios. These KPIs can be used for simulation scenario development to test particular sustainability aspects of ACT deployment and relevant policies.

Keywords: Connected and automated vehicles; traffic simulation; transport modelling; sustainability; key performance indicators; connected transport; autonomous vehicle

INTRODUCTION

Automated and connected transport (ACT), a.k.a. connected and autonomous vehicles (CAVs), might bring about a disruption with the potential to change people's lives in the coming years. It is expected that CAVs could largely change traffic safety and mobility patterns, and further enable mobility as barriers to driving will be alleviated. Meanwhile, CAVs could potentially open new business opportunities both for various transport service providers and ICT firms. In the case that some form of higher automation will be deployed in the following decades, driven both by necessity and by the current trends in the relevant industries and among stakeholders, the expected impacts are significant and multifaceted. It is important to be able to anticipate desirable and undesirable effects, via the appropriate definition of tools and methods that may be used to evaluate and assess such impacts on our transport systems, particularly those impacts related to sustainability.

Decision-makers use ACT simulation models to replicate CAV driving behaviour using real-life data to support them in their decision-making. However, due to the limited number of real-life trials, the data available to construct, calibrate and validate models is scarce. Many of the Autonomous Vehicle (AV) prototypes on public roads provide some insight into how the technology can be implemented in the real world. These prototypes are still limited by their very cautious driving styles, especially in Europe. Furthermore, the pilot projects are often limited to collective transport services, so-called shared AV (Guo et al., 2020; Narayanan et al., 2020). On the other hand, the USA and Singapore are examples of countries that have implemented ACT trials for individual vehicles, i.e. private cars. Still, these trials assume a cautious and risk-averse mode, where a driver is ready to intervene if the vehicle causes any danger. Hence, the technology is not yet ripe for thorough on-street testing, due to for example, but not limited to, the relatively low number of prototype vehicles currently on public roads and the manufacturers' proprietary constraints or confidentiality concerning the capabilities of said prototypes.

Therefore, developing and testing different ACT simulation scenarios is vital to analyse different impacts on sustainability as well as infrastructure requirements, road safety and travel behaviour. Simulations offer reduced costs compared to pilot projects and enable the study of additional complexity options, which may be difficult or too costly to replicate in real-world trials. Another important reason to exploit simulation scenarios rather than test trials lies in the geographical/spatial area that can be covered by the former, with relatively low resources. Indeed, in networks at the country level, it is impossible to muster the necessary number of prototype vehicles to carry out multi-assessment studies. This limitation can be easily overcome through the implementation of models and simulation scenarios.

It is important for the simulations and modelling studies to result in comparable and transferable results, especially when different geographical or boundary conditions are accounted for (Studer et al., 2019). KPIs are exploited based on common definitions. KPIs are essential for monitoring the performance of the transport system and support practitioners in effectively communicating the results to decision-makers, thus aiding improved and informed decision-making. This bridge between practitioners and decision-makers is a further reason for this chapter and the provided analysis on ACT KPIs focusing on sustainability. To improve our plans, the definition and monitoring of KPIs are essential to make improved decisions towards sustainability (Litman, 2007). KPIs are widely used to assess the performance of systems to meet stakeholder objectives in road infrastructure decision-making (Adey, 2019; Adey et al., 2019). An important aspect of the planning of infrastructure is the scenario development for an uncertain future. Often, the large uncertainty is simplified to best guesses or an expected value (McPhail et al., 2018). It is also worth stating what it is meant for scenario development throughout the chapter. ACT scenario development defines the scenario-building activities forecasting trends and uptakes of ACT technologies. As will be discussed in the next section, the literature already includes a wide range of scenarios with some conflicting trends and hypotheses, making scenario development a key step of the process.

This chapter aims to discuss the existing ACT simulation scenarios that have been tested worldwide (Dadashzadeh et al., 2021). This chapter focuses on capturing the large range of different studies related to simulations and subsequently defines KPIs quantifying impacts of future ACT scenarios. There are already reports and European projects focusing on KPIs for AVs (Agriesti et al., 2019; Innamaa & Kuisma, 2018; Penttinen et al., 2019; Studer et al., 2019), however, this chapter will be focused on simulation KPIs, related exclusively to sustainability (economic, social and environmental impacts) which are in the line with the UN Sustainable Development Goals (SDGs) Nos. 9, 11 and 13 (United Nations, 2015). The output of the said analysis is twofold: a comparison of the current literature concerning AV simulations and a set of criteria and KPIs exploitable in ACT simulation studies. The definition of the KPIs is important to investigate how and in which aspects could ACT deployment scenarios potentially impact the infrastructure (e.g. in terms of capacity utilisation and the need for new innovative infrastructure) (SDG No. 9), improve safety, and make cities liveable, accessible and sustainable (SDG No. 11). In addition, the identification of KPIs is required to quantify the impacts of different potential AV consumption technologies on air emissions and fuel savings (SDG No. 13).

This chapter is structured as follows. In the next section, it reviews the ACT-related research to summarise the aims and goals of different studies. This section will also present the main issues of simulation scenarios development and assessment. The following section presents the KPIs that can be used for developing future simulation scenarios considering sustainability aspects. Finally, the last section highlights the main findings and suggests future directions.

AN OVERVIEW OF SIMULATION STUDIES FOR AUTOMATED TRANSPORT

This section summarises the efforts within WISE-ACT WG5 COST Action (https://wise-act.eu/) to simulate ACT scenarios. This summary is expanded with a wider perspective from global activities related to ACT simulation scenarios. To have a comprehensive finding relevant to the topic, we reviewed a wide range of scientific literature and parallel projects. The objective of this section is to provide insights regarding existing simulation scenarios and KPIs. To this end, the summary table proposed by Dadashzadeh et al. (2021) has been modified to focus on ACT simulation scenarios assessing environmental, societal and economic impacts.

Throughout the chapter, we proceed by analysing the various aspects that have been considered within each of the following three key steps of a simulation pipeline: scenario building, simulation tools and performance implications.

- In this section, the aims and objectives of different studies investigating scenarios, how scenarios were defined and what assumptions are considered in shaping those scenarios are reported. In Table 3.1, we have included studies that include keywords such as sustainability, climate protection, emissions, congestion, safety, veh-km travelled, travel time, delay, fleet size and accessibility.
- Second, the models that are used to investigate the developed scenarios, comparing a contrasting methodological detail. The impact that the methods used may have on the result is further discussed to ensure a balanced comparison.
- Third, criteria and indicators that are typically reported as outcomes of simulations or other analyses are listed. These are then discussed in the context of reflecting on the performance of systems exposed to future ACT scenarios. This results in conclusions regarding the implications of the findings on policymaking.

SIMULATION PIPELINE, INPUT AND ASSUMPTIONS

Traffic simulation models are classified into micro, meso and macroscopic models. The characteristics of each model are distinguished based on their levels of complexity and network scale. Microscopic models are highly detailed, capturing

Table 3.1. Simulation Scenarios Related to ACT Studies (Adapted from Dadashzadeh et al., 2021).

Source	Country	Study Topic	Software	Model Type	Traffic Modes/ Composition Considered
Guériau and Dusparic (2020)	Ireland	Traffic efficiency, Traffic safety	Sumo	Microscopic	Mixed traffic[a]
The result of this study depicts the significant benefits of CAV deployment both on traffic efficiency and safety. Traffic efficiency improvements in congested situations are achieved a relatively lower ACT penetration rates. However, the exact penetration rate which brings such improvements depends on network type and traffic states and might be between 20% and 40%. The safety analysis shows that lower penetration rates result in a 30% increase in a number of conflicts, whereas with higher penetration rates, the number of conflicts reduces by around 50–80%.					
Emberger and Pfaffenbichler (2020)	Austria	Travel Behaviour	MARS (Metropolitan Activity Relocation Simulator – coded[b])	Land use and transport interaction model	90% of private AVs by the year 2050
The paper shows that privately operated AVs will have a large impact on veh-km travelled. The added comfort of this mode will lead to reductions in the use of active modes and public transport. Car dependence will increase. Without considerable changes to the fleet, the energy and resource consumption will also increase, as opposed to political goals towards net zero. However, introducing automation in the first and last mile, e.g. as part of public transport service, has the potential to reduce car dependence, veh-km travelled and environmental impacts.					
Nahmias-Biran et al. (2020)	Singapore	Social impact	SimMobility	Activity/Agent-based	Automated Mobility on Demand, Car, Carpooling, Bus, Mass Rapid Transit, Automated Taxi, Motorcycle and walking
The investigation of network-wide penetration of automated mobility on demand (AmoD) reveals significant accessibility and network performance. In terms of income, the high deployment of AmoD has a bigger influence on low-income individuals' accessibility compared to mid- and high-income individuals.					
Papantoniou et al. (2021)	Greece	Energy consumption and air emissions	SUMO	Microscopic	Mixed Traffic

(continues)

Table 3.1 (*Continued*)

Source	Country	Study Topic	Software	Model Type	Traffic Modes/ Composition Considered
The findings of this study show that any penetration rates of AVs could potentially increase the NO_x and CO emissions on motorways. This is shown for multiple scenarios for differing CAV market penetration.					
Agriesti et al. (2021)	Italy	Traffic Efficiency	PTV VISSIM	Microscopic	Mixed Traffic
The results of this study showed that when AVs are informed to conduct takeover manoeuvre in advance and before and bottleneck, the efficiency of the bottleneck significantly improves. This improvement depends on the distance which the cooperative message reaches to AVs. The entrance speed of 30 and 40 km/h during peak hour is achieved for penetration rates of 80 and 100% respectively with a range of C-ITS is 700 m. While this speed value is achieved at a 50% penetration rate if the C-ITC range is 1,500 m.					
Mattas et al. (2018)	Belgium	Energy consumption and air emissions	Aimsun	Microscopic	Mixed Traffic (both automation without connection and connected automated vehicles)
This study investigated the impacts of AVs and CAVs penetration rates on emissions per kilometre under different demand scales. The findings reveal that under lower demand, the fully human-driven vehicles' environment generates the highest emissions per km, and any penetration rates of AVs could reduce emissions. However, under higher demand scales and higher penetration rates of AVs, the emissions significantly increase. In comparison, CAVs improve the situation and reduce emissions.					
Hartmann et al. (2017)	Germany	Traffic Efficiency	PTV VISSIM	Microscopic	Mixed Traffic (German vehicle stock)
Considering the assumed conservative driving behaviour of AVs, the findings of this study show a negative impact on the capacity of freeways, whereas the automated technologies in AVs which result in shorter time gaps between vehicles, could potentially increase the freeway capacity by 30% and significantly reduce the traffic delays.					
Alam and Habib (2018)	Canada	Travel Behaviour	PTV VISSIM	Microscopic	Mixed Traffic

The study aimed to find the right fleet size. This is a trade-off between the ability to meet the demand in the peak hour and its utilisation. With increasing fleet size, the number of trips per shared vehicle decreases, but the ability to serve the trip requests for a morning peak hour commute improves.					
Dinar (2020)	Germany	Traffic Efficiency	PTV VISSIM	Microscopic	CAV, Car, Truck, Bike and Cargo Bike
The result of this study reveals that the best traffic efficiency improvement is achieved at 60%, 50% and 70% penetration rates under peak hour, 20% below peak hour and 20% above peak hour demands respectively. Optimum penetration levels are seen at the penetration of 60%, 50% and 70% for the peak hour demand, 20% below peak hour demand and 20% above peak hour demand, respectively. **The average travel time and delay time** of the study corridor decreased, while speed and connectivity, among C/AVs sharply increased, for high penetration rates of C/AV, up to the optimum penetration.					
Lu et al. (2021)	Germany	Traffic Efficiency	SUMO	Microscopic	Only AV
This research investigated the impacts of AVs with different driving styles (cautious, normal and aggressive) in a network. The findings depict that with 50% PR of AVs under different combinations of driving styles, the mean speed of the network drops and travel time increases. The combination of cautious and normal driving style (CN) can form the most stable network state and facilitate the 'delivery' efficiency of the network. Surprisingly, it is found that the worst **mean speed and travel time** are obtained at *r* (ratio between behaviours) = 0.5 for all combinations. Furthermore, in the experiments with all three driving styles, *C* is the least sensitive to the ratio, while *A* is the most sensitive.					
Guériau et al. (2020)	US	Traffic Efficiency	Sumo	Microscopic	Mixed Traffic
Travel time for particular trips between traffic zones is tested given different scenarios of different market penetration of shared automated vehicles.					
Sonnleitner et al. (2021)	Germany	Macroscopic assignment	PTV VISUM	Macroscopic	Car (driver and passenger) Mass Transit (PT), Walk and Bike
AVs were simulated by testing two assumptions: a linear relationship between the share of AV and their impact on capacity and a nonlinear relationship on the other hand. Different functions are tested to identify the relationship between the share and PCU of AVs. A 100% AV traffic is found to cause a 6% **increase in VMT**, considering the impact of AVs on perception time further raises the VMT upto 18% for an 80% AV share.					
Fernandes et al. (2021)	Portugal	Macroscopic assignment	PTV VISUM	Macroscopic	Car, Shared vehicles, Electric vehicles, AV, CAV

(*continues*)

Table 3.1 (*Continued*)

Source	Country	Study Topic	Software	Model Type	Traffic Modes/ Composition Considered
Different combinations of shared, electric and AVs have been assessed in their environmental impacts. AVs have been simulated in the macroscopic model through an increase of capacity factor equal to 1.1173 and to 1.1780 for a market penetration of 25 and 35% respectively. 13 scenarios are simulated, **CO_2 emissions** ranging between −22% and +6%, **NOx emissions** between −35% and +5%. The same emissions per vehicle range instead between −13% and 1% or −18% and ′+8% respectively.					
von Schmidt et al. (2022)	Germany	AV Impacts	SUMO	Mesoscopic/ Microscopic	AV, Car, Mass Transit, Walk and Bike
AVs were simulated at different levels (demand was also considered through ABM). Two sets of parameters were included to simulate urban areas as microscopic and other areas as mesoscopic. Three future scenarios are simulated, with AV shares ranging from 18 to 100%, the results suggest that the **mode share** for private vehicles may increase up to 5%, the **distance travelled** up to 9.3 million kilometres and a decrease in **travel time** up to 26% for long-distance trips and to 27% for commuters.					
Liu et al. (2017)	US	Travel Behaviour	MATSim	Agent-based/ Mesoscopic	AV, Car, Transit
Shared automated vehicles and their trip prices were simulated to understand the impact of price on demand, given a deterministic user equilibrium.					
Bösch et al. (2018)	Switzerland	Traffic Efficiency and travel behaviour	MATSim	Agent-based/ Mesoscopic	Mass Transit (PT), Walk, Bike, Car, Automated Taxi, Automated Public Transport
While increased automation may lead to positive effects through shared automated vehicles, there will also be a shift of users from private cars, public transit and slow modes to automated taxis. In combination with the increase in zero-occupancy rides, this will lead to additional VKT and increased risk of congestion.					
Hatzenbühler et al. (2020, 2021)	Sweden	Travel Behaviour	BusMezzo	Agent-based	Automated Buses, Mass Transit (PT)
Deploying automated buses leads to service frequency improvements and a reduction in passenger load per line on automated bus lines as well as a general shift from typical public transport modes to automated bus services. Since the service is available on demand, the number of served bus stops can be reduced. Additionally, the average passenger waiting time can be reduced when deploying automated buses instead of public transport further reducing the generalised costs borne by the user.					

[a] By mixed traffic it means any percentage of ACT coexisting on the same infrastructure with traditionally driven vehicles. No particular focus is given to other modes.

[b] By coded, it means a theoretical framework applied through coding rather than traditional software.

the driving behaviour of a single vehicle, which includes the car-following behaviour, lane change behaviour, and its interaction with other road users. In macroscopic modelling, instead, vehicles are considered as flow in a continuous stream, while flow, average velocity and density are the three main quantities considered in these models. They are less detailed and are utilised to investigate the assignment procedure using activity-based or modified four-step models. Finally, a mesoscopic model is a combination of macroscopic and microscopic models, which integrate the driving behaviour of each vehicle with the traffic flow features. They are simple but at the same time have reasonably high levels of detail (Treiber & Kesting, 2013).

This section investigates how the various simulation types have been adapted to consider ACT. Indeed, ACT presents different challenges at different simulation levels. Multiple studies have already tried to replicate some form of automated driving in a simulation environment. AIMSUN, SUMO and VISSIM have been used to evaluate, for example, traffic efficiency through speed measurements, capacity increases and flow stability at microscopic or mesoscopic level (Guériau et al., 2020; Hartmann et al., 2017; Mattas et al., 2018). Other common evaluations have focused on emissions at the vehicle level (Dinar, 2020; Eilbert et al., 2018) or on traffic safety (Arvin et al., 2021; Guériau & Dusparic, 2020; Studer et al., 2022). Microscopic simulations have also been applied to test different control logics for ACT (Lu et al., 2021). More rarely, a microscopic tool has been used to evaluate directly fleet sizes of shared autonomous vehicles (SAVs) (Alam & Habib, 2018). On the other side of the spectrum, studies considering ACT and using Agent-Based Modelling (ABM) have mostly focused on outputs such as demand distribution for different fare rates, service times for on-demand services and fleet sizes (Bösch et al., 2018; Hamadneh & Esztergár-Kiss, 2021; Liu et al., 2017). Other common research questions are related to occupancy rates of ACTs and willingness to share a ride (Ortega et al., 2020). Some overlap with mesoscopic studies, albeit at different scales, can still be found. In evaluating trip times or VKT, for example, the indicators assessed traffic efficiency through metrics such as travel time or delay. Still, while in the former set of tools, the changes in travel time resulted from changes in traffic flow dynamics, in the latter, the change was typically the result of a different number of vehicles on the roads.

Most importantly, one simulates how ACT disrupts different simulation pipelines in different ways. For example, micro and mesoscopic simulations rely heavily on hypotheses about control algorithms and prototype performances. These data are not made public by any major car manufacturer and, more importantly, the models used to replicate ACTs have not been extensively validated yet. This means that the uncertainty about how ACTs will or should interact with human drivers and vice-versa (both in real life and consequently in the model) is quite high. Regarding ABM on the other hand, the criticalities relate more to the demand and regulatory side. Indeed, while these simulations try to frame how individuals will choose ACTs based services and which current trips will change once these are available, only hypotheses may be made on two important aspects: How the willingness to use automated vehicles will change with time across different population segments and how regulators will adapt to account for the unavoidable conflicts between automated and non-automated modes. To make a very simple example, if a speed limit will be imposed on automated vehicles in the same way as it happened with e-scooters across many European countries, it should be clear how most of the current assessment should be adopted.

Fig. 3.1 provides an overview of the four main pipelines when it comes to simulations. These do not refer to a specific software but define, instead, the 'level' of the simulation experiments. Each level has its strengths and weaknesses but note that the aim of this section is not to rate them. Fig. 3.1 is based on the literature analysis carried out in the second section and summarised in Table 3.1; it focuses

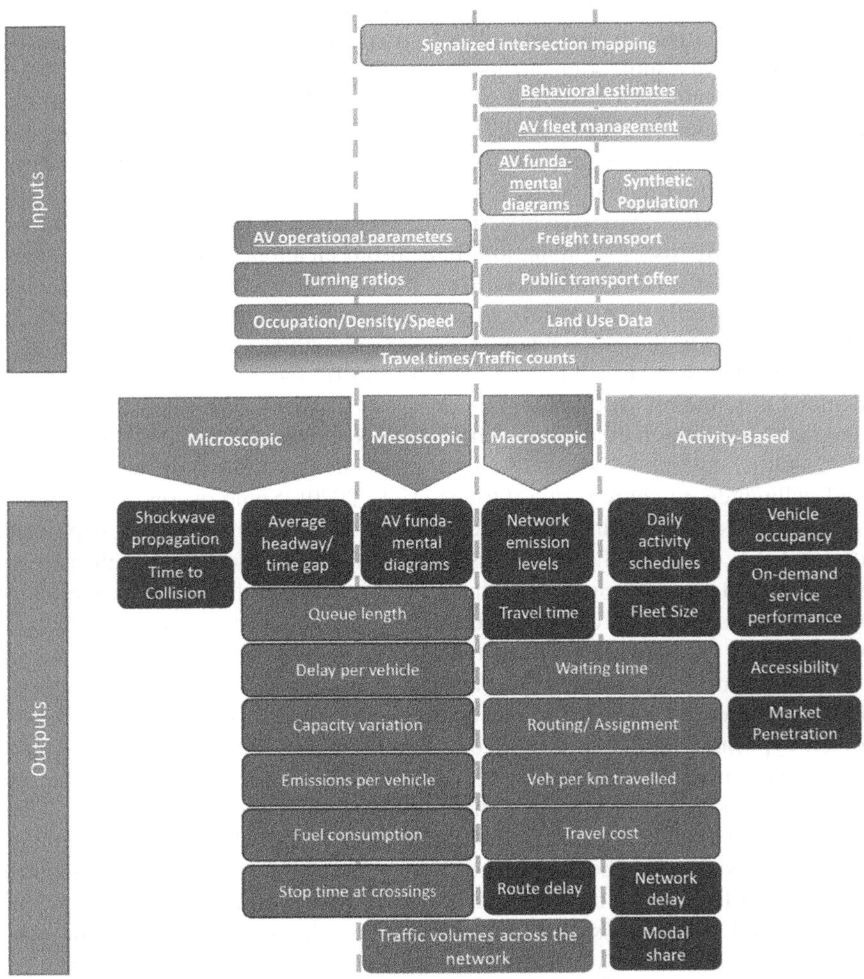

Fig. 3.1. Simulation Pipelines Considering ACT and Sustainability Based on Reviewed Literature in Table 3.1 (*Image Source*: Authors).

on outputs concerning sustainability and it is further commented on in the last section.

Fig. 3.1 highlights how ACT studies fit in the current simulation environment and ACT-specific items have been underlined. In the upper part, the required data and inputs for each level are listed. It is worth noting how microscopic and mesoscopic studies require only one ACT-specific item: operational parameters (i.e. headway/time gap, take-over time, lane change aggressiveness). Currently, the main microscopic/mesoscopic software has embedded some default values for these parameters [e.g. AIMSUN: (CoEXist, 2020), PTV VISSIM: (Zeidler et al., 2019)]. Still, it is worth emphasising how no fully automated vehicle is available on the market and how any of the current prototypes do not share information about these operational specifications. Thus, caution is required when choosing said parameters. On the other hand, macroscopic and activity-based models, require different inputs, the latter being particularly data-demanding. The same stands also for ACT, with modal share usually defined by relying both on behavioural modelling (activity-based) or scenario building (macroscopic). Other items not related to ACT modelling have been included to give a picture of the overall data requirement needed to build a simulation experiment for each of the four levels. In the figure, the term network means a large infrastructure and it is used to characterize large-scale outputs (e.g. emissions). In the bottom half, the most common indicators that may be obtained through simulations have been reported. While a more complete analysis (concerning ACT) will be provided in the following section, Fig. 3.1 ties the various KPIs to the corresponding simulation levels. It is worth highlighting, for example, how most of the KPIs related to (ACT) capacity or traffic efficiency result from micro and mesoscopic simulations. These are instead inputs for the other simulation levels. Socioeconomic or spatial outputs arise instead from macroscopic and activity-based models.

While Fig. 3.1 is based on the studies reported in Table 3.1 and the ones reported at the start of this section, a wider ACT literature is available concerning for example AV response time (Schoettle Brandon, 2017), driver characterisation and impact assessments. The studies concerning the latter and their focus on KPIs have also been used to build Fig. 3.1 (Agriesti et al., 2019; Innamaa & Kuisma, 2018; Penttinen et al., 2019; Studer et al., 2019, Dadashzadeh et al., 2021). Still, the figure clusters some of the categories in the literature for simplicity (e.g. ACT reaction times and driver characterisation are included as AV operational parameters). Most of the reported boxes may either be open in different KPIs (reported later) or already represent KPIs relevant to sustainability. These are emissions per vehicle, emissions at a network level, fuel consumption but also fleet size, modal share or accessibility.

The KPIs related to emissions have been differentiated to account for different kinds of impacts. Fuel consumption, while like emissions per vehicle, accounts for pollutants directly released from the vehicle, most likely in densely populated areas. Emissions per vehicle instead, as KPI reflects the pollution due to the energy needed for the vehicle to move. Still, with the advent of electric vehicles, this energy production and the resulting pollutants are essentially decoupled from the network and call for different kinds of assessments. Emissions at the network level reflect instead the overall emissions level across the whole network, without considering directly their source. More operational KPIs and outputs [e.g. ACT vehicle jerk, take over time (Alrefaie et al., 2019) or trajectory] have not been considered or included because they are outside the scope of the current analysis. Fig. 3.1 is intended as a map framing the general trends and capabilities of the various tools available, a more detailed analysis concerning outputs (and thus the lower half of the figure) is provided in the following section. Finally, since Fig. 3.1 does not refer to specific software or tools, the reported input and output clustering is general by nature and should be taken as a guiding tool, supporting the choice of the right model for the task at hand.

KPIS AND RELATION TO POLICY NEEDS

Efficient realisation of ACT deployment into existing transport infrastructure requires qualitative and quantitative assessments and predictions for the potential impacts. For this, simulation-based impact assessments are utilised. Simulation-based platforms act as important tools for evaluating new transport strategies, policies and innovations. Many studies have reported the potential effects of ACT deployment scenarios using traffic models. New trends in mobility will affect how stakeholders define their infrastructure needs. The stakeholder needs are expressed through the expected levels of service quantifications. The Output Indicators in Fig. 3.1 are used to evaluate the performance of the transport system and quantify environmental impacts, while also providing a proxy for the infrastructure performance. These are of high relevance for decision-makers as infrastructure owners, for example, road agencies, are largely exposed to the uncertainty regarding the possible deployment of new vehicle technology. If the new needs can be monitored closely, infrastructure interventions or developments can be introduced promptly for the effective management of infrastructure. Due to the longevity of infrastructure and high investment costs, infrastructure owners have a great interest in being aware of the uncertainty related to the deployment of future mobility alternatives and the wider impacts of these. To hedge the risk, it is important to plan for different futures exploring multiple scenarios and aligning these with stakeholder goals.

In each ACT simulation-based study, a set of assessment objectives, criteria and indicators have been selected to quantify the impact of ACT penetration. Among other key strategic areas, sustainability aspects of ACT impacts have been widely studied in the literature. These sustainability goals consist of objectives and KPIs for quantifying various scenarios of ACT deployment. Mainly accessibility, mobility pattern, traffic efficiency, operational efficiency, safety, energy consumption and air emissions are the key sustainable strategic areas that lead the policymakers to decide about the potential utilisation of ACT scenarios in different localities (Sadid & Antoniou, 2023). In concrete terms, ACT deployment scenarios could be assessed using simulations in the following impact fields. The relevant KPIs listed in Table 3.2 (based on the output indicators in Fig. 3.1) are used for assessing policy decisions and decision-making support, for example,

Table 3.2. List of KPIs Should be Considered for ACT Simulation Scenarios.

Strategic Area	Sub-category	KPIs	Unit
Traffic efficiency	Mobility	Average travel time per road km	minute
		Total network travel time	minute
		Maximum road capacity for a road section	veh/h
		Speed of a given road section	m/s or km/h
		Delay time	minute
	Reliability	Travel time variability	%
		Predictability of travel time	%
		Effective capacity	veh/h
		Speed distribution	Unitless
Safety	-	Number of conflicts	number
		Time to collision (TTC)	second
		Post-encroachment time (PET)	second
Energy consumption and air emissions	Energy consumption	Fuel consumption of a vehicle	litres/100 km
		Total fuel consumption per year	ton
		Electric energy consumption per vehicle	kWh/100 km
	Air pollutants	CO_2 emissions per vehicle-km per hour	g/km/hr
		Annual traffic CO_2 emissions on a route	tons/year
		NO_X, CO, PM_{10}, $PM_{2.5}$ and VOC emissions per vehicle km	mg/km
Travel behaviour	Modal split	Modal split per week	%
		Travel time by mode per week	hours
	Trip characteristics	Total kilometres travelled per week	km/week
		Number of trips per week	number
Land use	Parking areas	Number of parking slots	Parking units per km^2
	Housing density	Houses	Housing units per km^2

ensuring the sustainability impact of a given policy or assessing whether a policy is robust to achieving net-zero targets.

Traffic Efficiency

ACT deployment will provide the opportunity to efficiently manage the traffic flow by enabling the vehicles to adapt their driving behaviour and route choice to the existing situation. In addition, the ability of ACT to generate a stream of data using information and communication technology (ICT) will better coordinate the flow of traffic through intersections and consequently improve the capacity of the network. The term traffic efficiency covers a variety of aspects including link and network flow of traffic. Potential KPIs for simulation-based assessment of traffic efficiency are mobility and reliability-related measures. Mobility in this context defines how the transport system provides access to different activity regions including jobs, recreation, shopping and more. Quantifying the mobility reflects the potential impact of ACT in the network both in private and public transport. On the other hand, reliability indicates the variability of mobility indicators. The higher the level of reliability of the transport system, the better the efficiency of the traffic flow. The set of indicators associated with mobility and reliability is listed in Table 3.2.

Safety

ACT deployment into transport systems will significantly change the driving behaviour of road users and consequently the safety of the transport network. Various indicators could be used in simulation-based studies to quantify the safety level. The safety level is measured as the number of accidents in the network as well as the impacts on vehicles and other users. The simulation-based analysis uses safety surrogate measures to assess the safety of the network. The number and severity of conflicts (i.e. time to collision, post-encroachment time) are the most used indicators for the quantification of traffic safety.

Energy Consumption and Emissions

The transport sector is still responsible for a significant share of fossil fuel consumption worldwide, which leads to a substantial contribution in terms of air pollution. Automation and connectivity of vehicles, change in driving behaviours, as well as the electrification of ACT, can have significant impacts on energy consumption and air pollution. Most of this impact would be from the electrification of the ACT as well as SAVs. To get more detailed information in this regard, see Narayanan et al. (2020) who comprehensively reviewed the existing studies in particular simulation scenarios, to show the effectiveness of SAVs in reducing energy consumption and emissions. However, simulation-based studies consider the same vehicle technology in terms of combustion system or electric equivalent to investigate the impact of ACT driving behaviour on the change of energy consumption and air emissions. Assessment indicators range from total CO_2 emissions to pollutant emissions (i.e. NOX, CO, PM10, etc.).

Travel Behaviour

The travel behaviour of the travellers may change due to the ACT deployment. This includes changes in the selection of modes, routes and destinations which may cause the production of fewer trips. There will be a significant effect on personal mobility and travel behaviour due to the high level of automation. The most related KPIs for travel behaviour are related to modal split and type of trip by a traveller as shown in Table 3.2.

Land Use

Automation may have impacts on the functions of the land use. For instance, in long-term changes in land use, the availability and location of parking areas, as well as housing density, can play a vital role in the automation. In addition, with the high level of automation, the location of the pick-up and drop-off for Automated Taxi and accessibility to them become crucial elements.

CONCLUSIONS

This chapter contributes with an overview of simulations for ACT, highlighting the challenges and the opportunities, with a focus on evaluating its impact on sustainability. Firstly, studies that have focused on simulating the impact of ACT are summarised and organised to describe different evaluation scenarios. Then, the data requirements and achievable outputs are reported based on the considered simulation level. Lastly, KPIs are proposed to monitor the performance of the simulated transport systems and related to the policy needs.

It is worth noting that, currently, transport models and simulation tools are typically calibrated and validated through real data (e.g. vehicle speeds, headways, travellers' decisions, etc.). However, while ACT is yet to be deployed, the knowledge that practitioners and policymakers have is very limited, because any real data is based on current technology and early pilots, whereas future ACT systems may differ strongly from the current ones. The related challenges are only worsened by various unsolved questions relating to the sustainability impacts of ACT deployment. Different assumptions in developing simulation scenarios have been considered to explore these impacts. However, comparing different works and how the impacts lead to different modelling outcomes is not easy. This is because there are no universal methods or KPIs used to simulate ACT scenarios. While different modelling software (e.g. AIMSUN, VISSIM, SUMO, MATSim, etc.) are increasingly providing additional resources to improve the software' abilities to simulate ACT, the large uncertainty regarding their real-world performance stands in the way of calibrating and validating simulation models. Consequently, no qualified statement on the accuracy or usefulness of models to evaluate sustainability KPIs can be made. Identification of the KPIs and related impacts is very important for future policy building and recommendation.

Thus, making reasonable assumptions and designing appropriate scenarios becomes a key step in ACT assessment and evaluation. It is important to choose the right simulation tool and KPIs depending on the research question and the scope and scale of the analysis. This is especially true and difficult for ACT simulations, which will require a shift in the way of thinking about and using simulation tools. Simulation pipelines (Fig. 3.1), scenarios tested by different tools (Table 3.1) and a list of KPIs (Table 3.2) presented in this chapter provide practitioners, scholars and policymakers with the support to make the appropriate choice by identifying the most suitable modelling tools and KPIs among the ones at disposal. The chapter strives to summarise the related best practices and results through the lens of sustainability and to characterise the simulation process from start (scenario building) to finish (KPIs and assessment).

An appropriate set of simulated scenarios can assist decision-makers in better understanding the future mobility system and the impact of their decisions given new boundary constraints, be it automated connected transport, ride-shared services or any other technological deployments. For example, there is an open-source (demo system) simulation platform that practitioners and policymakers can use to test different ACT scenarios (Saigol, 2020). This chapter sheds light on the state-of-the-art simulation methods used for ACT, the challenges and opportunities for simulation tools due to ACT deployment and how to move forward with their application for meeting sustainability objectives such as the Net Zero targets and the UN Sustainable Development Goals.

ACKNOWLEDGEMENTS

This work was partly supported by the European Cooperation in Science and Technology (COST) Action: CA16222 titled 'Wider Impacts and Scenario Evaluation of Autonomous and Connected Transport' (WISE-ACT: https://wise-act.eu/).

REFERENCES

Adey, B. T. (2019). A road infrastructure asset management process: Gains in efficiency and effectiveness. *Infrastructure Asset Management*, *6*(1), 2–14. https://doi.org/10.1680/jinam.17.00018

Adey, B. T., Martani, C., Kielhauser, C., Urquijo, I. R., Papathanasiou, N., Burkhalter, M., & Beltran-Hernando, I. (2019). FORESEE – D1.1: Guideline to measure service provided by, and resilience of, transport infrastructure. In *8th Transport Research Arena: Rethinking transport. Towards clean and inclusive mobility (TRA 2020)*. ETH Zurich, Institute of Construction and Infrastructure Management. https://doi.org/10.3929/ethz-b-000403298

Agriesti, S., Studer, L. P., Ponti, M., Gandini, P., Marchionni, G., Borghetti, F., Paglino, V., Maja, R., & Bruglieri, M. (2019). Evaluation approach for a combined implementation of Day 1 C-ITS and highway chauffeur. In *2018 IEEE 87th Vehicular Technology Conference (VTC Spring)* (pp. 1–6). IEEE.

Agriesti, S., Ponti, M., Marchionni, G., & Gandini, P. (2021). Cooperative messages to enhance the performance of L3 vehicles approaching roadworks. *European Transport Research Review*, *13*(1), 1–19. https://doi.org/10.1186/s12544-020-00457-z

Alam, Md J., & Habib, M. A. (2018). Investigation of the impacts of shared autonomous vehicle operation in Halifax, Canada using a dynamic traffic microsimulation model. *Procedia Computer Science*, *130*, 496–503. https://doi.org/10.1016/j.procs.2018.04.066

Alrefaie, M. T., Summerskill, S., & Jackon, T. W. (2019). In a heartbeat: Using driver's physiological changes to determine the quality of a takeover in highly automated vehicles. *Accident Analysis & Prevention*, *131*, 180–190.

Arvin, R., Khattak, A. J., Kamrani, M., & Rio-Torres, J. (2021). Safety evaluation of connected and automated vehicles in mixed traffic with conventional vehicles at intersections. *Journal of Intelligent Transportation Systems*, *25*(2), 170–187.

Brandon S. (2017). *Sensor fusion: A comparison of sensing capabilities of human drivers and highly automated vehicles*. Report No. SWT-2017-12. The University of Michigan.

Bösch, P. M., Ciari, F., & Axhausen, K. W. (2018). Transport policy optimization with autonomous vehicles. *Transportation Research Record*, *2672*(8), 698–707. https://doi.org/10.1177/0361198118791391

CoEXist. (2020). Working towards a Shared Road Network: Enabling Cities to Get "Automation-Ready."

Dadashzadeh, N., Elvarsson, A., Morshed, G., Agriesti, S., Antoniou, C., Roncoli, C., & Thomopoulos, N. (2021). *Autonomous and connected transport scenarios evaluation based on simulation analysis*: WG5 Thematic Report.

Dinar, Y. (2020). *Impact of connected and/or autonomous vehicles in mixed traffic*. Technical University of Munich.

Eilbert, A., Lauren, J., George, N., & Scott, S. (2018). A framework for evaluating energy and emissions impacts of connected and automated vehicles through traffic microsimulations. *Transportation Research Board 97th Annual Meeting Proceedings*. https://trid.trb.org/view/1497138

Emberger, G., & Pfaffenbichler, P. (2020). A quantitative analysis of potential impacts of automated vehicles in Austria using a dynamic integrated land use and transport interaction model. *Transport Policy*, *98*, 57–67. https://doi.org/10.1016/j.tranpol.2020.06.014

Fernandes, P., Bandeira, J. M., & Coelho, M. C. (2021). A macroscopic approach for assessing the environmental performance of shared, automated, electric mobility in an intercity corridor. *Journal of Intelligent Transportation Systems*, 1–17. https://doi.org/10.1080/15472450.2021.1945453

Guo, J., Susilo, Y., Antoniou, C., & Brenden, A. P. (2020). Influence of individual perceptions on the decision to adopt automated bus services. *Sustainability (Switzerland)*, *12*(16), 1–13. https://doi.org/10.3390/su12166484

Guériau, M., Cugurullo, F., Acheampong, R. A., & Dusparic, I. (2020). Shared autonomous mobility on demand: A learning-based approach and its performance in the presence of traffic congestion. *IEEE Intelligent Transportation Systems Magazine*, *12*(4), 208–218. https://doi.org/10.1109/MITS.2020.3014417

Guériau, M., & Dusparic, I. (2020). Quantifying the impact of connected and autonomous vehicles on traffic efficiency and safety in mixed traffic. In 2020 IEEE 23rd International Conference on Intelligent Transportation Systems, ITSC 2020 (pp. 1–8). https://doi.org/10.1109/ITSC45102.2020.9294174

Hamadneh, J., & Esztergár-Kiss, D. (2021). Potential travel time reduction with autonomous vehicles for different types of travellers. *Promet – Traffic & Transportation*, *33*(1), 61–76. https://doi.org/10.7307/ptt.v33i1.3585

Hartmann, M., Krause, S., Hoffmann, S., Motamedidehkordi, N., Vortisch, P., & Busch, F. (2017, November). *Impact of automated vehicles on capacity of the German freeway network VISSIM view project virtual reality and pedestrian simulation view project impact of automated vehicles on capacity of the German freeway network*.

Hatzenbühler, J., Cats, O., & Jenelius, E. (2020). Transitioning towards the deployment of line-based autonomous buses: Consequences for service frequency and vehicle capacity. *Transportation Research Part A: Policy and Practice*, *138*, 491–507. https://doi.org/10.1016/j.tra.2020.06.019

Hatzenbühler, J., Cats, O., & Jenelius, E. (2021). Network design for line-based autonomous bus services. *Transportation*, *49*, 467–502. https://doi.org/10.1007/s11116-021-10183-7

Innamaa, S., & Kuisma, S. (2018). Key Performance Indicators for Assessing the Impacts of Automation in Road Transportation: Results of the Trilateral Key Performance Indicator Survey.

Litman, T. (2007). Developing indicators for comprehensive and sustainable transport planning. *Transportation Research Record*, 10–15. https://doi.org/10.3141/2017-02

Liu, J., Kockelman, K. M., Boesch, P. M., & Ciari, F. (2017). Tracking a system of shared autonomous vehicles across the Austin, Texas network using agent-based simulation. *Transportation*, *44*(6), 1261–1278. https://doi.org/10.1007/s11116-017-9811-1

Lu, Q.-L., Qurashi, M., Varesanovic, D., Sodnik, J., & Antoniou, C. (2021). Exploring the influence of automated driving styles on network efficiency. *Transportation Research Procedia*, *52*, 380–387. https://doi.org/10.1016/j.trpro.2021.01.094

Mattas, K., Makridis, M., Hallac, P., Raposo, M. A., Thiel, C., Toledo, T., & Ciuffo, B. (2018). Simulating deployment of connectivity and automation on the antwerp ring road. *IET Intelligent Transport Systems*, *12*(9), 1036–1044. https://doi.org/10.1049/iet-its.2018.5287

McPhail, C., Maier, H. R., Kwakkel, J. H., Giuliani, M., Castelletti, A., & Westra, S. (2018). Robustness metrics: How are they calculated, when should they be used and why do they give different results? *Earth's Future*, *6*(2), 169–191. https://doi.org/10.1002/2017EF000649

Nahmias-Biran, B. H., Oke, J. B., Kumar, N., Azevedo, C. L., & Ben-Akiva, M. (2020). Evaluating the impacts of shared automated mobility on-demand services: An activity-based accessibility approach. *Transportation*, (0123456789). https://doi.org/10.1007/s11116-020-10106-y

Narayanan, S., Chaniotakis, E., & Antoniou, C. (2020). Shared autonomous vehicle services: A comprehensive review. *Transportation Research Part C: Emerging Technologies*, *111*, 255–293. https://doi.org/10.1016/j.trc.2019.12.008

Ortega, J., Hamadneh, J., Esztergár-Kiss, D., & Tóth, J. (2020). Simulation of the daily activity plans of travelers using the park-and-ride system and autonomous vehicles: Work and shopping trip purposes. *Applied Sciences (Switzerland)*, *10*(8). https://doi.org/10.3390/APP10082912

Papantoniou, P., Kalliga, V., & Antoniou, C. (2021). How autonomous vehicles may affect vehicle emissions on motorways. In I. K. Eftihia, G. Nathanail, & G. Adamos (Eds.), *Advances in mobility-as-a-service systems* (Vol. 73, pp. 12–13). Springer International Publishing.

Penttinen, M., van der Tuin, M., Farah, H., de Almeida Correia, G. H., Wadud, Z., Carsten, O., & Kulmala, R. (2019). Deliverable D3.1: Impacts of connected and automated vehicles – state of the art. In *Conference of European Directors of Roads Project Nr. 867448 (MANTRA)*. https://www.cedr.eu/download/D3.1-Impacts-of-connected-and-automated-vehicles-State-of-the-art.pdf

Sadid, H., & Antoniou, C. (2023). Modelling and simulation of (connected) autonomous vehicles longitudinal driving behavior: A state-of-the-art. *IET Intelligent Transportation Systems 2023*, 1–21. https://doi.org/10.1049/itr2.12337

Saigol, Z. (2020). *WISE-ACT WG5 Webinar: Multiuser scenario catalogue for connected and autonomous vehicles (MUSICC) project*. UK Catapult Connected Place. https://www.youtube.com/watch?v=tQs9RlsCsfU&t=1s

Sonnleitner, J., Friedrich, M., & Richter, E. (2021). Impacts of highly automated vehicles on travel demand: Macroscopic modeling methods and some results. *Transportation*, *49*, 927–950.

Studer, L., Agriesti, S., Gandini, P., Ponti, M., & Bischoff, S. (2019). Impact assessment of cooperative and automated vehicles. In *Cooperative intelligent transport systems: Towards high-level automated driving* (pp. 2–4). IET The Institution of Engineering and Technology.

Studer, L., Ponti, M., Agriesti, S., Sala, M., Gandini, P., Marchionni, G., & Gugiatti, E. (2022). Modeling analysis of automated and connected cars in signalized intersections. *International Journal of Intelligent Transportation Systems Research*, *20*(2), 379–397. https://doi.org/10.1007/s13177-021-00279-w

Zeidler, V., Buck, S. H., Kautzsch, L., & Vortisch, P. (2019). Simulation of autonomous vehicles based on Wiedemann's car following model in PTV vissim. In *98th Annual Meeting Transportation Research Board (TRB 2019)*, Washington, DC, USA, 13.01.2019 – 17.01.2019. https://www.researchgate.net/publication/331020495_Simulation_of_Autonomous_Vehicles_Based_on_Wiedemann's_Car_Following_Model_in_PTV_Vissim

CHAPTER 4

APPLYING SERIOUS GAMES IN MODELS OF PREFERENCES OF SHARED AUTOMATED VEHICLES

Shelly Etzioni[a], Mor Collins[b], Eran Ben-Elia[b] and Yoram Shiftan[a]

[a]*Faculty of Civil and Environmental Engineering, The Technion, Haifa, Israel*
[b]*Department of Geography and Environmental Development, Ben-Gurion University of the Negev, Beer Sheva, Israel*

ABSTRACT

Serious games (SGs) are virtual systems that allow the reconstruction of the laws governing the behavior of complex adaptive systems such as urban transportation and social interaction. Unlike stated preference-based studies, improved visualization, feedback, and scores mediate players' learning through experience. SG's potential to understand users' preferences regarding shared automated vehicles (SAVs) is developed. The investigation focused on three innovative, entirely automated commuting options: shared rides, shared cars, and automated transit. The research involved 10 participants actively involved in a competitive mode selection exercise, which emulated 50 workdays and was conducted in 10 separate sessions. The players aimed to maximize their overall score influenced by their mode choice, punctuality, and the other players' choices. SG-obtained data was used to estimate a game-based discrete choice model. The sustainability policy implications of game-based methods on the future adoption of SAVs and impacts on other modes are further discussed.

Keywords: Shared automated vehicles (SAVs); game-based models; serious games (SGs); ridesharing; sustainable mobility

Sustainable Automated and Connected Transport
Transport and Sustainability, Volume 19, 65–82

ISSN: 2044-9941/doi:10.1108/S2044-994120240000019004

INTRODUCTION

Objectives and Context

Shared automated vehicles (SAVs) could well mitigate congestion and reduce the market share of privately used automated vehicles (AVs) and cars in general. SAVs have the potential to replace the need for owning private vehicles, are more spatially and temporally flexible compared to public transport, ultimately will be more affordable than conventional private cars, may reduce road accidents, reduce the cost of travel, and have the potential to increase the mobility of immobile members of the society (Fagnant & Kockelman, 2015; Litman, 2018; Milakis et al., 2017, 2020; OECD, 2015; Van Acker et al., 2021). In this respect, it is important to differentiate between synchronous (i.e., ridesharing) or asynchronous (i.e., carsharing) vehicle sharing, as the energy and traffic consequences are quite different (Wadud et al., 2016) and are discussed later on in this chapter.

The debate between scholars presenting optimistic scenarios regarding these potentials versus the tremendous challenges stresses the importance of studying the factors influencing SAV choice and the policy guidelines to promote them and mitigate private AV ownership sustainably. Policymakers and transportation planners need to forecast the introduction of AVs as a crucial step to the success and sustainability of this emerging technology. In this chapter, we discuss the potential of a novel and emerging methodology – SGs – to understand the formation of users' preferences regarding the adoption of SAVs and their possible impacts on the sustainable use of other transportation modes.

This chapter is organized as follows. First, we review prevalent methodologies applied to study automated vehicle acceptance and user preferences, including SGs. The next section presents the methods and procedures used in this research, specifically the SG design, which allows us to model how virtually experiencing different travel alternatives and their individual and collective consequences influence mode choice, preference stabilization, and willingness to ride-sharing. The results of an estimated game-based discrete choice model, presented in the following section, provide insight into how experience, feedback, and social interaction shape preferences over time and how they can facilitate the use of sustainable modes, including SAVs. Finally, the last section discusses the main findings and their possible sustainability policy implications.

METHODS APPLIED IN AUTOMATED VEHICLE ACCEPTANCE AND USER PREFERENCE STUDIES

Generally, the research field of user preference, particularly AVs acceptance, has advanced well in recent years. When studying users' acceptance or modeling future modal market shares of yet unavailable modes, it is important to address possible behavioral biases. Namely, the hypothetical bias of "saying is not the same as doing" (Fifer et al., 2014). Yet, exploring the tradeoffs between different attributes, individual characteristics, and attitudes that influence AV mode choice is essential for policymakers and transportation planners to plan for an uncertain future.

Scholars devised, therefore, different methods to recreate the automated driving experience to deal with users lack of experience. The following sub-sections describe existing methods for studying AV acceptance and user preferences.

Revealed Preference Surveys

Revealed preferences (RP) is a well-known method to evaluate users' preferences and evaluate transportation innovations. A few RP studies were used on user preferences of automated buses. A survey among Toulouse pilot shuttle users showed that about 98% found it comfortable, 79% felt safe using the shuttle, and 91% said they would have boarded even if there was no operator on board (Ainsalu et al., 2018). In La Rochelle, six automated shuttles with a capacity of 10 passengers operated for 4.5 months. All had the same 1.4-km route, with 6 stops. About two-thirds of the passengers considered using automated public transport over a conventional one (Piao et al., 2016). As more and more automated services are pioneering the market, more and more RP studies will likely be published (Paddeu et al., 2020).

Stated Preference-Based Studies

As RP is impractical when studying nonexistent modes, SP is commonly used to investigate hypothetical states (Etzioni et al., 2020, 2021; Klein & Ben-Elia, 2016; Van Acker et al., 2021). In fact, according to Becker and Axhausen (2017), the most prevalent research method used to study AV acceptance is SP with online surveys. SP can contribute to meeting sustainability objectives when testing for hypothetical, new or non-existent services such as SAV and to test links with specific sustainability goals of SAVs.

The reliability of the data that choice experiments provide is limited and cannot be used to draw a conclusive picture regarding future states. Nonetheless, choice experiments provide an efficient measure to assess consumers' evaluation of multi-attributed decisions (Yang et al., 2009).

SP or stated choice generally involves the same technique. Usually, respondents choose their preferred alternative from each set of alternatives, described using a hypothetical profile containing several attributes, according to a specific design, with varied trip attributes such as cost, travel time, wait time, convenience, or comfort. The number of scenarios presented is determined based on the trade-off between two incentives: Reducing the required sample size by showing each respondent as many scenarios as possible. However, too many scenarios presented may cause the respondents to adopt simplifying decision heuristics (Arentze et al., 2003). While SP methods in transportation are commonly used and widely accepted, they often get criticized. Respondents tend to adopt coping strategies consistent with how they usually process similar information in realistic decisions (Hensher, 2006). SP is, naturally, not based on real choices; hence the results may be biased and not reflect realistic behavior. Uncertainty, current expectations, and perceived risks can influence choice (Winter et al., 2017). Hypothetical bias can result in over or underestimation of respondents' responses since the decisions are not concrete (Klein & Ben-Elia, 2016). Furthermore, this methodology may not apply to less literate individuals (Arentze et al., 2003). Of course, SP studies must be planned carefully, as the complexity level or design dimensionality may affect the quality of the data obtained and its validity (Arentze et al., 2003; Hensher, 2006).

Recently, attempts have been made to improve the reliability of SP studies using a Virtual Reality (VR) experience. VR could holistically capture the total commuting experience in automated modes (Bowman & McMahan, 2007; Jetter et al., 2020). Farouq et al. (2018) researched pedestrian preferences regarding AV and associated infrastructure changes in urban streets in Montreal using VR and compared it with text-only and visual aid SPs. They showed that VR provided a better understanding of the scenario and yielded more consistent results. Nonetheless, VR studies in transportation are still nascent, while the lack of repeated interaction and feedback between VR and choice remains a drawback.

Public Opinion Surveys

Several qualitative opinion surveys regarding AVs included data from multiple countries, both Western and non-western. Schoettle and Sivak (2014) distributed an extensive survey across China, India, Japan, the USA, the UK, and Australia, exploring the different attitudes toward AVs. Many individuals expressed concern regarding safety and control issues such as driving without driver control or AVs driving unoccupied. The results indicated that people in India and China (84%) had more positive views about AVs than the Japanese (43%) and were the only groups willing to pay extra for this technology. Furthermore, they expressed more interest (76% of Chinese respondents and 80% of the Indian respondents) in purchasing an AV compared with Japanese respondents (41%) (Schoettle & Sivak, 2014). Kyriakidis et al. (2015), surveyed over 5,000 consumers from 109 countries, examining user acceptance, concerns, and willingness to buy an AV with different automation levels. The main concerns indicated were safety, software hacking, and legal issues.

A survey conducted by Continental AG (Sommer, 2013) in Germany, China, Japan, and the US found that AV awareness was not equally represented. For example, Germans (67%) and Chinese (64%) were more aware of the newest developments relative to the Japanese (29%) or Americans (50%).

Simulators

Most studies used SP surveys to assess drivers' acceptability and willingness to use AVs. However, as driving simulators have high validity (Farah et al., 2008; Lee, Cameron, & Lee, 2003), some studies used a driving simulator to better understand driver behavior. For example, in a study from 2015, Participants were distributed into two groups – simple and elaborate practice. In contrast to the simple practice, the elaborate practice had manual control recovery and overtaking maneuvers. Reaction time was related to higher trust but only in the simple practice group. The results emphasize the importance of proper training to mitigate the negative consequences of machine overtrust (Payre et al., 2015). Another study which focused on machine trust found that reported trust after a simulated driving task was an essential component of AV acceptance (Molnar et al., 2018). Furthermore, Buckley et al. (2018) found that attitudes, subjective norms, and perceived behavioral control were significant predictors of intentions to use AV.

Naturalistic Experiments

Another method that is used to project passengers into an AV experience is naturalistic experiments. Harb et al. (2018) aimed to mimic private ownership of an automated vehicle by providing 13 subjects with free 60 hours of chauffeur service for exclusive household use within one week in San Francisco. The researchers tracked the participants' travel behavior for three weeks using an app that continuously records all trips, one week before, one week during, and one week after the experiment. An overall increase of 83% in total vehicle miles traveled (VMT) was observed. Furthermore, the proportion of empty trips was substantial, i.e., sending the driver on errands. However, travel behavior in the first week may not reflect passengers' long-term behavioral changes as the excitement of trying a new service may result in more trips. In reality, transportation services are not expected to be cost-free. Paddeu et al. (2021) used an SP task before and after a brief exposure in a SAV. The experience influenced people's mode choice preferences, shifting from car (the most preferred mode before the experiment) to AV-taxi and SAV-taxi (after the experiment). Paddeu et al. (2020) showed in a similar setting how comfort and trust ratings are affected by specific attributes of the ride experience. Fifty-five participants rode in an SAV shuttle at a test facility. A strong correlation was found between comfort and trust, which indicated trust is an important predictor of the perceived comfort of AVs. Naturally, such experiments can only be conducted on a small scale and suffer from a lack of realism. In this regard, we introduce the idea of using SGs as a method that may have substantial potential.

Serious Games

SGs are games that are not designed for the sake of pleasure. A SG is a digital environment in which participants participate in simulated conflicts defined by specific game rules and constraints (Tekinbas & Zimmerman, 2003). SGs are distinct from agent-based simulations, as the behavior of the participant is not dictated by software code but is instead shaped by the decisions and actions of a human player. SGs provide a versatile research setting, distilling the travel experience into tangible choices, and enabling the analysis of human travel behaviors within a dynamic and lifelike context. Multi-player games allow social interaction, and the players' actions can portray the social consequences. Repeated choice promotes implicit learning, facilitating cooperation and shaping preferences over time.

The design, planning, and validity of SG must be carefully considered. A SG is based on psychological, structural, process, and predictive validities (Peters et al., 1998). Psychological validity necessitates a game environment that facilitates natural, real-life player behavior. Structural validity demands that the different circumstances of reality, such as agents, rules, data, information, and norms, must be reflected in the game environment. Process validity requires that the processes that occur in real life, including flows of information or resources, interactions between agents, and negotiations, should also be reflected in the game environment. Predictive validity requires that the game be predictive of real-life outcomes or be able to reproduce a historic outcome. The researcher must account for the validity of the game in all four criteria, thus preventing errors in the translation of reality into the game setting and game results back into reality.

The utilization of SGs for investigating decision-making within the transportation sector remains relatively rare. However, it is expected that such research will gain momentum due to the increasing prevalence of Information Communication Technologies (ICTs) and the accessibility of Advanced Travel Information Services (ATIS), which enable travelers to engage in online communication and access travel-related information. Klein and Ben-Elia (2017) used a route choice game on a simple road network to study how cooperation can reduce congestion. In their study, travelers could interact and were provided with basic travel information. In one condition, players were instructed to follow prescriptive information. In contrast, in the other condition, they were also punished for not complying with the information and rewarded if they did comply. Results demonstrated that punishments and rewards were more effective than only perspective information and brought the transportation system closer to system optimum (Klein & Ben-Elia, 2017).

More recently, The PARKGAMESG was developed to study the parking choice dynamics in urban areas (Fulman et al., 2020). A SG was applied to study two intertwined parking choices on a Manhattan grid road network: where to cruise for parking and when to quit. Players faced a day-to-day dilemma where they had to arrive on time to a meeting and decide whether to look for cheap but uncertain on-street parking or choose a certain costly lot parking. Being late for the meeting resulted in penalties. Results showed drivers were more risk-averse when parking and quit looking for street parking too early. Geva and Ben-Elia (2022) expanded the SG to include pricing. They showed that the interaction between price and occupancy influenced parking choices. Higher rates reduced the motivation to cruise, whereas lower rates motivated more cruising further away from the destination.

To the best of our knowledge, no study used SGs to study mode choice behavior in general or specifically AVs. With the advent of AVs enabled by V2V ICTs, the potential of SGs for automated mode choice studies seems vast.

RESEARCH METHODS

We present a SG-based study of asynchronous and synchronous SAVs acceptance. Ten participants jointly faced a computerized mode choice social dilemma where they were asked to choose their preferred mode out of the three alternatives for their daily commute. The proposed game setting was used to study users' choice and adoption of three novel and fully automated transportation modes: Private ride in AV (SC), Shared ride in AV (SR), and Automated public transit (PT). One hundred players were recruited. The players aimed to maximize their overall payoff, which was affected by their mode choice, punctuality, and other players' choices. Each game had 50 rounds; a number was found sufficient for behavioral convergence and trend stabilization but not too long to induce the player's fatigue and loss of attention span.

The players were students recruited using the ORSEE web-based Online Recruitment System (Greiner, 2015). The experiment had 10 sessions. Due to the COVID-19 outbreak and social distancing regulations, the experiment was conducted online through May–July 2020, using the cloud. As the operational parameters were calibrated before any COVID-19-related concerns and restrictions, the players were asked to consider their travel patterns under this assumption.

The participants were initially required to complete a brief survey encompassing socio-demographic details and questions related to their travel habits. Subsequently, they engaged in a mode choice game. After the survey, they were presented with instructions for the game.

In this simulated daily commute, participants were tasked with traveling from their homes to their workplace or college. Each day was assumed to commence at 8:00 AM, and their objective was to arrive by 9:30 AM to optimize their daily score or performance in each round. It was emphasized that arriving either too late or too early would result in a reduction in their overall payoff. Importantly, the penalty for tardiness was more severe than that for arriving early, as being excessively early was categorized as idle time, whereas being late had an adverse impact on work productivity.

The participants were required to choose their preferred commuting mode from among the three available options (SR, SC, and PT) each morning. Upon completing this task, each player received compensation ranging from 40 to 100 Israeli Shekel (ILS), contingent on their performance in comparison to their fellow participants.

The participants were provided with the following descriptions of the three alternatives (originally in Hebrew):

1. **Private ride in an automated vehicle (SC):** In this scenario, you have exclusive access to a self-driving vehicle for your entire journey. It can transport you to your destination at your preferred time. However, if numerous players opt for this choice, it could lead to congestion, subsequently increasing travel times.
2. **Shared ride in an autonomous car (SR):** With this option, you share your ride with one other passenger and split the cost, but the travel time may be extended due to the pickup or drop-off of the additional passenger.
3. **Automated public transit (PT):** This service operates on a fixed schedule, departing precisely at 8:30 AM. The travel duration remains consistent at one hour, regardless of the number of players who select this option. It follows a designated route.

The operational parameters of the three modes were designed to reflect different levels of risk-taking, with SC being the riskiest alternative, SR intermediate-risk, and PT, being risk-neutral. A predominantly risk-averse player would choose PT which offered a low but certain score throughout the game.

A choice screen was presented following the instructions and the description of the modes (Fig. 4.1). This screen presented each choice with a time-slider to indicate the chosen departure time. The presented order of the three modes was randomized in each round to maintain players' attention and prevent any biases associated with the position of the alternatives. The first five rounds were training rounds. Subsequently, two indicators (calculated as moving average since the beginning of the game) were added to the choice screen: (1) The average expected travel time by road (only relevant for SC and SR) and (2) The cumulative average number of passengers choosing SR. These indicators were created to support the social-dilemma nature of the game and to simulate real-life conditions where information on the expected travel time and market shares are available and inherently dependent on the mode choices of other travelers (see Fig. 4.1). A feedback screen was presented (Fig. 4.2) after the choice screen.

On the feedback screen, participants were presented with the results of their choices, which included the real travel time and the corresponding score for the particular round. Additionally, the anticipated travel time displayed on the choice screen was included as a reference point.

Subsequently, the participants proceeded to the next round of the game. The players were not informed that the game had 50 rounds to avoid any strategic behaviors associated with the number of rounds (the shadow of the future). The experiment lasted about 40 minutes.

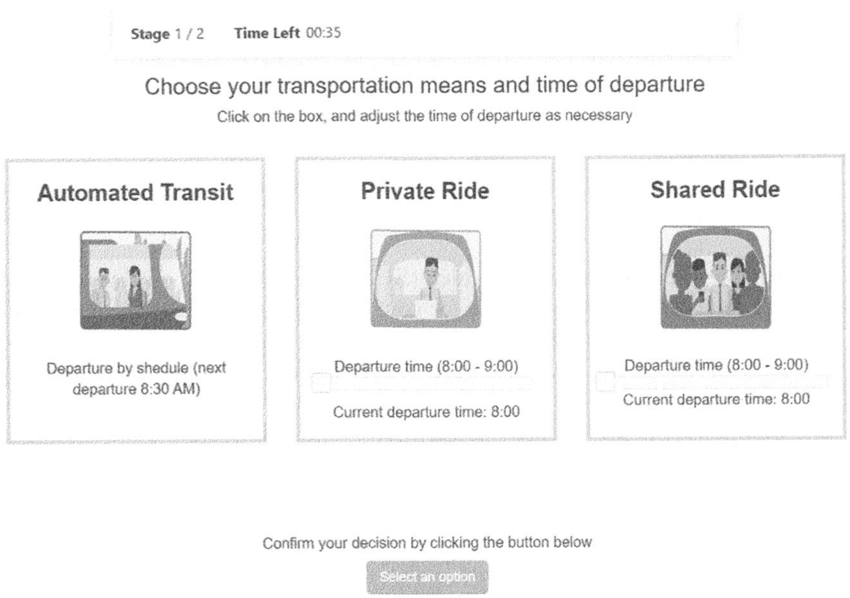

Fig. 4.1. Screenshot of the Mode Choice Game (After the 5th Round).

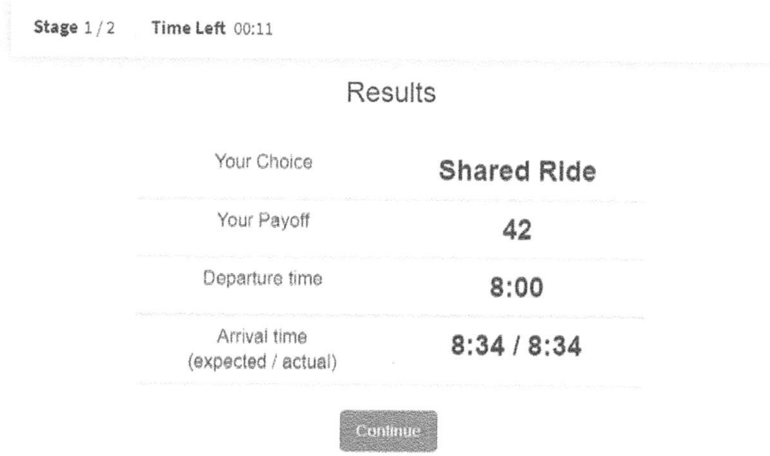

Fig. 4.2. Screenshot of the Feedback Screen.

CHOICE MODEL

A game-based model (GBM) was estimated to understand user preferences while elucidating how experience, feedback, and social interaction shape preferences over time and how they can facilitate the use of more sustainable transportation modes. A mixed-multinomial logit (MMNL) model, known as the logit kernel (LK), was specified with cross-nested error components and panel data. The specification provided the following properties:

- Experience – The repeated game nature allowed us to model how an individual's experience with the different alternatives and their social consequences influence choice and preference formation.
- Travel time – Travel time directly generated by the choices of the other players was used in the GBM, unlike conventional SP studies where individual choice is not affected by the choices of other system agents.
- Sharing – The Game-based data was used to model the effect of the social dilemma introduced by the game on the willingness to share.
- Departure time – Framing the commute experience to a simplified game whereby arriving late or too early results in penalties that provide insights into how players adjust their departure time to follow an optimal schedule, depending on the chosen alternative.

Choice Probabilities

A total of 100 participants, randomly distributed across 10 game sessions, each consisting of 10 players, collectively made 5,000 mode choices through 50 simulated "daily" decision points. In these choices, each player had to select both the mode of transportation and their departure time. When considering the mode preferences alone, the reported shares across all 50 rounds were as follows: 59% for Shared-Ride (SR), 30% for Shared Car (SC), and 11% for Public Transit (PT). Although SR emerged as the most favored mode, the distribution of departure times indicated that the majority of players preferred to start their journeys within the last 20 minutes, with the average departure time occurring at the 43rd minute.

Specification

To capture the repetitive nature of the game choices and the complex choice structure encompassing both departure time and mode, a cross-nested LK-type model was selected, utilizing panel data. Within this model, a random parameter (RNP) was introduced to account for unobservable taste variations that persist over time, as suggested by Ben-Akiva et al. (2001) and Walker et al. (2007).

The model also featured a cross-nesting structure facilitated by five pseudo-cross-nesting parameters, termed error components (CN). These parameters included two to capture correlations across different modes and three to address different departure times corresponding to early, intermediate, and late departures within SR or SC. In a cross-nesting framework, the alternatives can belong to multiple groups, as outlined by Walker (2001). For instance, "Early SR" would belong to both the "SR" group and the "early departure" group.

The model estimation was carried out using Pandas Biogeme (Bierlaire, 2020). In addition to the cross-nesting structure and the RNP, five other variables were integrated into the final models: expected travel time, the consequence of being late, the status of being an experienced player (after the 20th round), and the daily score.

The departure times for SR and SC, initially recorded as a continuous variable ranging from 1 to 60 minutes, were redefined as a categorical variable comprising three departure time slots: Early, Intermediate, or Late. This categorization aimed to align the departure times with the penalties associated with early or late arrivals. Importantly, empirical examination of the models indicated that the departure times naturally clustered into these three time slots. Departures made within the first 38 minutes were designated as Early, those occurring between the 39th and 43rd minute were classified as Intermediate (Mid), and departures after the 43rd minute were categorized as Late. This classification was based on the distribution of the departure time data, with approximately one-third of departures falling into each time slot. In summary, the model encompassed seven alternatives based on the chosen mode and departure time.

RESULTS AND INFERENCE

All parameters exhibit statistical significance at the 0.05% level, with the exception of the constants associated with Early SR and Mid SR, which were found to be statistically insignificant, as indicated in Table 4.1.

Table 4.1. Game-Based Model.

Parameter	SR			SC			PT
	Early	Mid	Late	Early	Mid	Late	–
Constant	−0.852 (−1.01)	−0.249 (−1.26)	2.880 (5.40)	−0.852(−2.28)	−1.040 (−2.74)	2.880 (10.7)	0-fixed
RNP intercept	0.811 (15.40)						
Expected travel time	−0.219 (−3.11)	–	−0.508 (−8.59)	−0.284 (−4.40)		−0.711 (−17.00)	0-fixed
Experience	–	1.250 (9.37)	—	–	0.771 (5.64)		0-fixed
Late	0.835 (1.99)	–	–	0.961 (3.68)			0-fixed
Avg shared	1.120 (3.83)	1.780 (6.70)	–	–	1.520 (4.25)	0.980 (3.55)	0-fixed
Score	0-fixed	3.970 (5.89)	–	−3.350 (−2.23)	–	–	−3.120 (−3.36)
CN Early	0.801 (5.41)	–	–	0.801 (5.41)	–	–	–
C.N. Late			1.260 (8.50)			1.260 (8.50)	
C.N. Mid		1.140 (7.76)			1.140 (7.76)		
CN SC				0.304 (2.18)			
CN SR	0.346 (2.74)				–		
Sample size	100						
observations	4,500						
Excluded observations	500						
Initial loglikelihood	−4929						
Final loglikelihood	−4703						
Draws	2,500						

Notes: "–" non-significant parameter that was fixed to 0.00.
T-tests are shown in parenthesis.

- The significant **RNP Intercept**, which captures the correlation across time for different agents, affirms that players' taste heterogeneity persisted over time, aligning with expectations.
- The negative sign of **Expected Travel Time** suggests that longer SR/SC travel times had a negative impact on the players' utility. Notably, the effect was more pronounced for late departures, indicating that longer expected travel times encouraged earlier departures. This parameter remained consistent in SC for both the Early and Mid categories, as a log-likelihood ratio test showed no statistical distinctions.
- The positive and significant **Experience** parameters for Mid/Late SC and Mid SR imply that experienced players exhibited a preference for these alternatives (see Fig. 4.3). This led to higher scores and supported the notion that experience with new modes can influence behavioral changes. A likelihood ratio test further confirmed the absence of statistical differences between the Mid and Late SC parameters.
- An increase in **AVG Share**, signifying that more players chose to share, corresponded to a greater probability of opting for Mid/Early SR. This supports the hypothesis that social interaction encouraged sharing, as a higher number of players preferred Mid/Early departures to avoid late arrival penalties. The likelihood of selecting Mid/Late SC also increased, with Mid departures being more favored than Late departures. This outcome can be attributed to the increased travel time resulting from the SR detour penalty parameter, which steered players toward SC when the demand for SR exceeded a certain threshold.
- **Arriving Late** in the previous round augmented the probability of choosing Early SR and SC (Early, Mid, and Late) in the subsequent round. Being late heightened the likelihood of selecting earlier departures in SR to avoid late arrival penalties and also increased the likelihood of choosing SC.
- A positive Score parameter for Mid SR and a negative one for Early SC and PT indicated that a higher score in the preceding round prompted players to opt for Mid SR and discouraged the selection of PT or Early SC in the following round. This outcome reinforced the idea that players implicitly learned which modes led to the highest scores through the feedback system and adjusted their choices accordingly.
- Positive **Cross-nesting (CN)** parameters confirmed the presence of an unobservable correlation among different modes and departure times. Notably, SR choices and Late departures exhibited the strongest correlation.

Mode and Departure Time Choice Dynamics

The mode choice patterns in each round are presented in Fig. 4.3. The shares began to stabilize for all modes after the 16th round. While the PT's share decreased after the trial rounds, the SR share increased, indicating that experienced players learned to choose SR. After the trial rounds, the SC choice trend remains more or less stable.

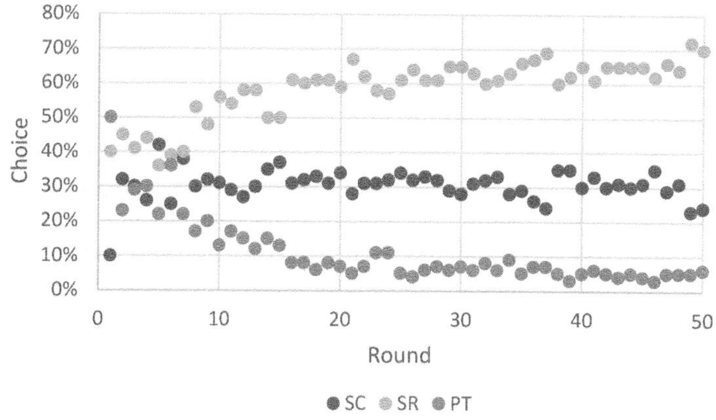

Fig. 4.3. Mode Choice Proportions by Round.

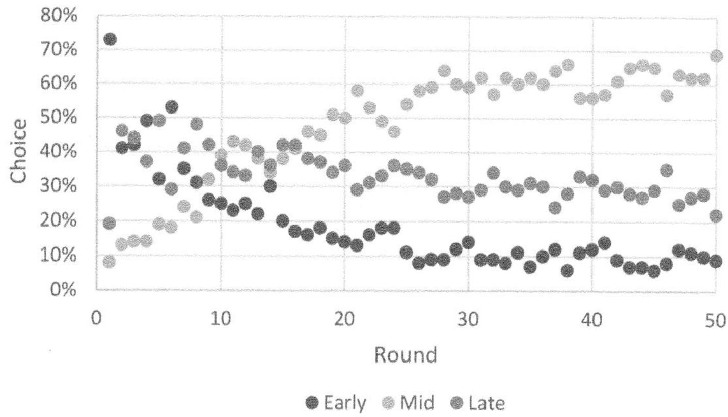

Fig. 4.4. Departure Time Choice Proportions by Round.

Fig. 4.4 shows the departure time choice patterns as a function of experience. The trends begin to stabilize after the 10th round. The Mid departure shares steadily increased with experience, indicating that the players learned that Mid departures result in higher scores. Late and Early departure shares decreased, but the decrease in the Early departure shares was sharper.

DISCUSSION, POLICY IMPLICATIONS, AND CONCLUSIONS

In this chapter, we reviewed the existing methods involved in understanding the formation of travelers' preferences toward SAVs and their limitations. We showed how an emerging methodology based on SG and GBMs, has the potential to better understand the preference of nonexistent modes like SAVs while accounting for the problem of hypothetical bias which afflicts SP surveys.

The way that AVs and SAVs are introduced is crucial to the technology's acceptance or rejection by customers and its impact on transportation sustainability goals, especially as the market is likely heading toward a digitally integrated Mobility-as-a-Service (MaaS) operational model (Hensher et al., 2021; Wong et al., 2020). The SG setting allowed us to investigate the dynamics of experience and feedback that cannot be studied using traditional SP. GBMs allow modeling complementary dynamics such as experience and social interaction and choices of other agents on mode choice that cannot be accounted for with survey-based data. The effects of social interactions and experience were modeled in a cross-nesting LK model. The model results showed choice trends stabilized over time, with an increasing preference toward ridesharing (59%), mainly at the expense of PT and, to some extent, on SC.

This result is unsurprising given the inherent advantages of a temporally flexible SR over a fixed schedule PT relative to a more costly and risky SC option. From a smart and sustainable mobility policy viewpoint, on the positive side, SR can become quite attractive, at least from a temporal perspective, thus reducing the necessity to own an AV in the future and considering that SR will provide a sufficient level of service. Naturally, a reduction in private rides will have positive environmental externalities, including pollutants and Green House Gases (GHGs) from the transport sector. However, a reduction in the attractiveness of PT has negative consequences and can also have equity impacts on socially vulnerable groups. This disturbing result is already evident in North American cities where ridesharing displaces PT services, already negatively affected by COVID-19 social distancing (Loa et al., 2021). In this respect, one noteworthy limitation is that our study ignored spatial effects. However, as shown in Ben-Dor et al. (2019), spatial heterogeneity and operational constraints (e.g., maximal allowed detour distance and waiting times) result in inadequate SR services, leaving travelers in metropolitan peripheries highly dependent on PT. In this sense, future SAV services must be carefully planned and regulated to avoid market failures.

The results obtained through the GBM analysis illustrate how the dissemination of information and the encouragement of sharing influenced other players to transition to SR, adjust their departure times, and ultimately set off a chain reaction that had a positive impact on road traffic conditions. As more players opted for SR, the likelihood of selecting it with an early or intermediate departure increased. Given that ATIS and related applications are already widely utilized for trip planning, such as Moovit (Haque et al., 2013), these findings offer valuable insights for policymakers. They highlight the potential of leveraging multimodal information in policymaking efforts to promote ride-sharing and manage its demand effectively, ultimately contributing to more efficient and sustainable transportation systems.

Several limitations are important to note. First, the SG considered only time-related factors. In reality, travelers will make decisions based not only on time but also on travel costs. SAVs will likely be more costly than PT services while still cheaper than private rides or owning an AV. Therefore, future work should also include variable costs in simulation and models. Second, as noted above, spatial effects should also be considered. Third, comfort issues also play a role and could be introduced in game design. For example, Etzioni et al. (2021) showed negative impacts of seating arrangements on SR preference. These can be integrated into game and simulation design (Muller et al., 2021). In addition, health-related issues are also an issue of concern, as evident in the avoidance of shared modes and PT during the COVID-19 lockdowns (Hensher, 2020). Privacy issues, riding with strangers, unwanted conversations, or unpleasant odors can also negatively impact sharing preferences. In this respect, a possible future extension will be integrating SGs with maturing VR technology to create a more holistic, immersive, and emotionally challenging experience (Anthes et al., 2016; Jetter et al., 2020; Klein & Ben-Elia, 2016), far beyond the current state-of-the-art.

ACKNOWLEDGMENTS

Funding was provided by the Israel Ministry of Transportation and Road Safety research project "Adoption of the Shared Autonomous Vehicle: A Game-Based Model of Urban Travelers and Transportation System Co-Adaptation" (Grant Cycle 9/2016); the Institute for Transportation Innovation at Tel Aviv University research project "Game-Based Modeling of Societal Transition to MaaS" Grant Cycle 2018; Action CA16222 of the European Cooperation in Science and Technology (COST) "Wider Impacts and Scenario Evaluation of Autonomous and Connected Transport" (WISE-ACT); and the Israel Science Foundation (Grant 1554/17) "Activity and Travel Behavior Implications of Autonomous Vehicles Using a Naturalistic Experiment." The first author is thankful for the support of the Technion-Israel Institute of Technology. The first and second authors are grateful to the Israel Smart Transportation Research Center (ISTRC) student scholarships.

REFERENCES

Ainsalu, J., Arffman, V., Bellone, I. D. M., Ellner, M., Haapamäki, T., Vaskinn, I., Haavisto, N., Josefson, E., Ismailogullari, A., Lee, B., Madland, O., Madžulis, R., Müür, J., Mäkinen, S., Nousiainen, V., Pilli-Sihvola, E., Rutanen, E., Sahala, S., Schønfeldt, B., Smolnicki, P. M., Soe, R.-M., Sääski, J., Szymańska, M., Vaskinn, I., & Åman, M. (2018). State of the art of automated buses. *Sustainability*, *10*, 1–34. https://doi.org/10.20944/preprints201807.0218.v2

Anthes, C., García-Hernández, R. J., Wiedemann, M., & Kranzlmüller, D. (2016, March). State of the art of virtual reality technology. In *2016 IEEE aerospace conference* (pp. 1–19). IEEE.

Arentze, T., Borgers, A., Timmermans, H., & DelMistro, R. (2003). Transport stated choice responses: Effects of task complexity, presentation format and literacy. *Transportation Research Part E: Logistics and Transportation Review*, *39*(3), 229–244. https://doi.org/10.1016/S1366-5545(02)00047-9

Becker, F., & Axhausen, K. W. (2017). Literature review on surveys investigating the acceptance of automated vehicles. *Transportation*, *44*(6), 1293–1306. https://doi.org/10.1007/s11116-017-9808-9

Ben-Akiva, M., Bolduc, D., & Walker, J. (2001). Specification, identification and estimation of the logit kernel (or continuous mixed logit) model. *Department of Civil* ..., 1–61. http://emlab.berkeley.edu/reprints/misc/multinomial2.pdf

Ben-Dor, G., Ben-Elia, E., & Benenson, I. (2019). Determining an optimal fleet size for a reliable shared automated vehicle ride-sharing service. *Procedia Computer Science*, *151*, 878–883.

Bierlaire, M. (2020). *A short introduction to PandasBiogeme*.

Bowman, D. A., & McMahan, R. P. (2007). Virtual reality: How much immersion is enough? *Computer*, *40*(7), 36–43. https://doi.org/10.1109/MC.2007.257

Buckley, L., Kaye, S. A., & Pradhan, A. K. (2018). Psychosocial factors associated with intended use of automated vehicles: A simulated driving study. *Accident Analysis and Prevention*, *115*, 202–208. https://doi.org/10.1016/j.aap.2018.03.021

Etzioni, S., Daziano, R. A., Ben-Elia, E., & Shiftan, Y. (2021). Preferences for shared automated vehicles: A hybrid latent class modeling approach. *Transportation Research Part C: Emerging Technologies*, *125*, 103013.

Etzioni, S., Hamadneh, J., Elvarsson, A. B., Esztergár-Kiss, D., Djukanovic, M., Neophytou, S. N., ... & Shiftan, Y. (2020). Modeling cross-national differences in automated vehicle acceptance. *Sustainability*, *12*(22), 9765.

Fagnant, D. J., & Kockelman, K. (2015). Preparing a nation for autonomous vehicles: Opportunities, barriers and policy recommendations. *Transportation Research Part A: Policy and Practice*, *77*, 167–181.

Farah, H., Yechiam, E., Bekhor, S., Toledo, T., & Polus, A. (2008). Association of risk proneness in overtaking maneuvers with impaired decision making. *Transportation Research Part F: Traffic Psychology and Behaviour*, *11*(5), 313–323.

Farooq, B., Cherchi, E., & Sobhani, A. (2018). Virtual immersive reality for stated preference travel behavior experiments: A case study of autonomous vehicles on urban roads. *Transportation research record*, *2672*(50), 35–45.

Fifer, S., Rose, J., & Greaves, S. (2014). Hypothetical bias in stated choice experiments: Is it a problem? And if so, how do we deal with it? *Transportation Research Part A: Policy and Practice*, *61*, 164–177. https://doi.org/10.1016/j.tra.2013.12.010

Fulman, N., Benenson, I., & Ben Elia, E. (2020). Modeling parking search behavior in the city center: A game-based approach. *Transportation Research Part C: Emerging Technologies*, *120*, 102800. https://doi.org/10.1016/j.trc.2020.102800

Geva, S., & Ben-Elia, E. (2022, March). Cruising Drivers' Response to Changes in Parking Prices in a Serious Game. In *Advances in Social Simulation: Proceedings of the 16th Social Simulation Conference*, (pp. 175–189). Cham: Springer International Publishing.

Greiner, B. (2015). Subject pool recruitment procedures: Organizing experiments with ORSEE. *Journal of the Economic Science Association*, *1*(1), 114–125. https://doi.org/10.1007/s40881-015-0004-4

Haque, M. M., Chin, H. C., & Debnath, A. K. (2013). Sustainable, safe, smart-three key elements of Singapore's evolving transport policies. *Transport Policy*, *27*, 20–31. https://doi.org/10.1016/j.tranpol.2012.11.017

Harb, M., Xiano, Y., Circella, G., Mokhtarian, P. L., & Walker, J. L. (2018). Projecting travelers into a world of self-driving vehicles: Estimating travel behavior implications via a naturalistic experiment. *Transportation*, *45*, 1–17.

Hensher, D. A. (2006). How do respondents process stated choice experiments? Attribute consideration under varying information load. *Journal of Applied Econometrics*, *47*(4), 36–37. https://doi.org/10.1002/jae

Hensher, D. A. (2020). What might Covid-19 mean for mobility as a service (MaaS)?. *Transport Reviews*, *40*(5), 551–556.

Hensher, D. A., Mulley, C., & Nelson, J. D. (2021). Mobility as a service (MaaS)–Going somewhere or nowhere? *Transport Policy*, *111*, 153–156.

Jetter, H.-C., Rädle, R., Feuchtner, T., Anthes, C., Friedl, J., & Klokmose, C. N. (2020). In VR, everything is possible!: Sketching and simulating spatially-aware interactive spaces in virtual reality. In *Proceedings of the 2020 CHI conference on human factors in computing systems* (pp. 1–16). Association for Computing Machinery.

Klein, I., & Ben-Elia, E. (2016). Emergence of cooperation in congested road networks using ICT and future and emerging technologies: A game-based review. *Transportation Research Part C: Emerging Technologies*, *72*, 10–28. https://doi.org/10.1016/j.trc.2016.09.005

Klein, I. & Ben-Elia, E. (2017). System optimal ATIS as a congestion management instrument – Game-based experiment and agent based model. In *2017 5th IEEE International Conference on Models and Technologies for Intelligent Transportation Systems (Mt-Its)*, (701), pp. 816–820.

Kyriakidis, M., Happee, R., & de Winter, J. C. (2015). Public opinion on automated driving: Results of an international questionnaire among 5000 respondents. *Transportation research part F: traffic psychology and behaviour, 32,* 127–140.

Lee, H. C., Cameron, D., & Lee, A. H. (2003). Assessing the driving performance of older adult drivers: On-road versus simulated driving. *Accident Analysis & Prevention, 35*(5), 797–803.

Litman, T. (2018). Autonomous vehicle implementation predictions: Implications for transport planning. *Transportation Research Board Annual Meeting, 4,* 36–42. https://doi.org/10.1613/jair.301

Loa, P., Hossain, S., Mashrur, S. M., Liu, Y., Wang, K., Ong, F., & Habib, K. N. (2021). Exploring the impacts of the COVID-19 pandemic on modality profiles for non-mandatory trips in the Greater Toronto Area. *Transport Policy, 110,* 71–85.

Milakis, D., Thomopoulos, N., & Van Wee, B. (2020). *Policy implications of autonomous vehicles.* Academic Press.

Milakis, D., van Arem, B., & van Wee, B. (2017). Policy and society related implications of automated driving: A review of literature and directions for future research. *Journal of Intelligent Transportation Systems, 21*(4), 1–25. https://doi.org/10.1080/15472450.2017.1291351

Molnar, L. J., Ryan, L. H., Pradhan, A. K., Eby, D. W., St. Louis, R. M., & Zakrajsek, J. S. (2018). Understanding trust and acceptance of automated vehicles: An exploratory simulator study of transfer of control between automated and manual driving. *Transportation Research Part F: Traffic Psychology and Behaviour, 58,* 319–328. https://doi.org/10.1016/j.trf.2018.06.004

Muller, M., Park, S., Lee, R., Fusco, B., & Correia, G. H. D. A. (2021). Review of whole system simulation methodologies for assessing mobility as a service (Maas) as an enabler for sustainable urban mobility. *Sustainability, 13*(10), 5591.

OECD. (2015). Urban mobility system upgrade: How shared self-driving cars could change city traffic. *Corporate Partnership Board Report,* 1–36. https://doi.org/10.1007/s10273-016-2048-3

Paddeu, D., Shergold, I., & Parkhurst, G. (2020). The social perspective on policy towards local shared autonomous vehicle services (LSAVS). *Transport Policy, 98,* 116–126.

Paddeu, D., Tsouros, I., Parkhurst, G., Polydoropoulou, A., & Shergold, I. (2021). A study of users' preferences after a brief exposure in a Shared Autonomous Vehicle (SAV). *Transportation Research Procedia, 52,* 533–540.

Payre, W., Cestac, J., & Delhomme, P. (2015). Fully automated driving: Impact of trust and practice on manual control recovery. *Human Factors: The Journal of the Human Factors and Ergonomics Society, 58*(2), 229–241. https://doi.org/10.1177/0018720815612319

Peters, V., Vissers, G., & Heijne, G. (1998). The validity of games. *Simulation & Gaming, 29*(1), 20–30.

Piao, J., McDonald, M., Hounsell, N., Graindorge, M., Graindorge, T., & Malhene, N. (2016). Public views towards implementation of automated vehicles in urban areas. *Transportation Research Procedia, 14*(0), 2168–2177. https://doi.org/10.1016/j.trpro.2016.05.232

Schoettle, B., & Sivak, M. (2014). *A survey of public opinion about autonomous and self-driving vehicles in the U.S., the U.K., and Australia* (pp. 1–85). University of Michigan, Ann Arbor, Transportation Research Institute.

Tekinbas, K. S., & Zimmerman, E. (2003). *Rules of play: Game design fundamentals.* MIT Press.

van Acker, V., Kyriakidis, M., Thomopoulos, N., Kolarova, V., Etzioni, S., Baburajan, V., Pagoni, I., Kouniadi, D., Pronello, C., Pourhashem, G., Shiftan, Y., Polydoropoulou, A., & Abreu e Silva, J. (2021). *WG4 thematic report: Transport demand & user challenges of ACT, COST Action CA16222 WISE ACT.*

Wadud, Z., MacKenzie, D., & Leiby, P. (2016). Help or hindrance? The travel, energy and carbon impacts of highly automated vehicles. *Transportation Research Part A: Policy and Practice, 86,* 1–18.

Walker, J. L. (2001). *Extended discrete choice models: Integrated framework, flexible error structures, and latent variables extended discrete choice models: Integrated framework, flexible error structures, and latent variables.*

Walker, J. L., Ben-Akiva, M., & Bolduc, D. (2007). Identification of parameters in normal error component logit-mixture (NECLM) models. *Journal of Applied Econometrics*, *1125*, 1095–1125. https://doi.org/10.1002/jae

Winter, K., Oded, C., Martens, K., & van Arem, B. (2017). Stated choice experiment on mode choice in an era of free-floating carsharing and shared autonomous vehicles: Raw data. *Transportation Research Board, 96th Annual Meeting*, (1), 1–17. https://doi.org/10.4121/uuid:4ac4d7b7-c8b0-42ec-a096-55a4f1837585

Wong, Y. Z., Hensher, D. A., & Mulley, C. (2020). Mobility as a service (MaaS): Charting a future context. *Transportation Research Part A: Policy and Practice*, *131*, 5–19.

Yang, L., Choudhury, C. F., Ben-Akiva, M., Abreu e Silva, J., & Carvalho, D. (2009). *Stated preference survey for new smart transport modes and services: Design, Pilot study and new revision*. Lisbon, Portugal: Massachusetts Institute of Technology.

PART TWO

POLICY & REGULATION

CHAPTER 5

CAN DRIVERLESS TRANSPORT BE SUSTAINABLE? A TRIPLE BOTTOM LINE PROBLEMATISATION

Alexandros Nikitas

Huddersfield Business School, University of Huddersfield, Queensgate, Huddersfield, UK

ABSTRACT

Connected and autonomous mobility may be an imminent game-changing reality, still in its embryonic form, that is set to disrupt a century-long 'driver-centric' status quo and recalibrate transport in unprecedented and possibly entirely unexpected ways. Autonomous vehicles (AVs) may have among others, a major impact on sustainability which in an era where concerns about the urgency and magnitude of climate change threats are voiced more and louder than ever before, needs to be a positive one for helping societies to enjoy liveable futures. This might not be an easy task to accomplish, however. This chapter, using a thematically organised narrative review approach, tries to give a well-rounded answer on whether driverless technology can yield sustainability benefits (or not) by looking at all three spheres of sustainability referring to environmental, economic and social implications. Agendas like motor traffic, air pollution, energy consumption, employment dynamics, inclusion, cybersecurity and privacy are all explored, and a conclusion is derived highlighting the need to package automation with connectivity, alternative fuelling and multimodality and building it around public transport (and to a lesser extent sharing service) provision. The road to make driverless transport genuinely sustainable is 'bumpy' and 'uphill' and requires the development of an appetite not for technology excellence per se, but rather for travel behaviour change. Achieving

Sustainable Automated and Connected Transport
Transport and Sustainability, Volume 19, 85–100

ISSN: 2044-9941/doi:10.1108/S2044-994120240000019005

this needs serious strategic and coordinated multi-stakeholder efforts in terms of pro-active policy reform, user (and transport provider) education and training initiatives, infrastructure investment, business plan development, and living lab experimentation.

Keywords: Automated transport; connected and autonomous vehicles; automated mobility; sustainability; liveable driverless futures; sustainable mobility

INTRODUCTION

Vehicle automation carries an unprecedented transformative potential for urban passenger transport (Thomopoulos & Givoni, 2015) and has critical sustainability implications for the future of our planet, eco-systems, cities and livelihoods (Nikitas et al., 2021). A mobility paradigm centred around automated transport and in particular around Connected and Autonomous Vehicles (CAVs) can be revolutionary but also disruptive (Nikitas et al., 2019) in all facets, expressions and proportions of the diverse and multidimensional concept of sustainability (Toli & Murtagh, 2020). This is a concept that in light of the rapidly growing climate change threats that we finally are taking seriously, has transformed to the 'Holy Grail' of urban planning in general and transport planning in particular (Nikitas, 2018).

On the one hand, automation is a powerful AI-centric apparatus that under the right conditions may improve: road accident prevention rates; riding experience; vehicle-sharing business models; road traffic congestion alleviation; accessibility and social inclusion for those currently unable to drive or have their own private vehicle; environmental degradation avoidance; energy saving rates; and our response to air pollution problems (Fagnant & Kockelman, 2015; Nikitas et al., 2021). On the other hand though, transport automation if used irresponsibly, in a fragmented manner and without a consistent strategic plan may instead lead to increased vulnerability to hacking, software and hardware flaws; privacy loss and travel data exploitation; liability allocation challenges; higher demand for car travel and elevated traffic accident and congestion rates especially in mixed traffic situations due to increased solo travel; behavioural adaptation, situational awareness, and user resistance problems; adverse labour market disruption and expensive skill reform schemes; and a rise in emissions and resource overconsumption (Liu et al., 2020; Nikitas et al., 2021).

This chapter attempts to provide a holistic answer on whether AV technology for multimodal city travel in general and CAVs, in particular, can yield sustainability benefits (or not) by looking at all three spheres of sustainability referring to environmental, economic and social considerations. The work thoroughly discusses some crucial potential impacts and side-effects of AV inclusion in the travel eco-system that reflect and affect motor traffic, air pollution, energy consumption, employment, inclusion, privacy and cybersecurity among others.

The chapter then identifies and contextualises the importance of packaging automation with symbiotic functions referring to connectivity, alternative fuelling and multimodality highlighting the fact that automated mobility should be principally built around public transport operations disincentivising the use of car-centric services.

Henceforth, the chapter provides in the next section a description of the research methodology. The following section looks into the three pillars of sustainability as organised by the triple bottom line theory. This approach allows this work to explore a multitude of aspects like motor traffic, air pollution, energy consumption, employment dynamics, accessibility, cybersecurity and privacy. This is followed by a solution-centred discussion that synthesises all the diverse dimensions of AVs and sustainability and projects the only foreseeable pathway for yielding positive outcomes from the transition to a driverless mobility paradigm. The last section offers conclusions and highlights future research directions that could expand the findings of this research work.

METHODOLOGY

This work is a thematically organised narrative review-based discussion. It uses peer-reviewed literature as an apparatus for building a critical and original problematisation about the key determinants of sustainability that may underpin automated mobility. The focus of the chapter goes beyond pro-environmental considerations per se and negotiates the equally important but typically under-appreciated forms, dimensions and expressions of sustainability aligned with the bottom line pillars of environment, society and economy. The work can therefore be classified as a literature review-enhanced position study that provides original insights about the uneasy alignment of AV-centric technology with genuine sustainability goals because of their possibly contradictory roles and priorities. This will hopefully support future transport planning efforts around balancing tensions and achieving beneficial sustainability outcomes when facilitating an eventual AV-centric paradigm transition. The chapter is a tool for contextualising the CAV deployment governance and its possible sustainability preconditions and safeguards meaning to ensure that an automated transport eco-system could lead to improved wellbeing.

More specifically, the literature review approach that is the basis of the chapter identified and analysed Scopus-listed journal papers written in English to ensure a minimum level of quality, while limiting the number of papers to be reviewed in the case of high-volume research fields such as automated mobility. The choice of the references was made, based on the subject-specific relevance and recency of each research output, the respective journal impact factor and each paper's impact as measured by the number of its citations (as a proxy for 'some older' papers). This is a tested methodological approach in line with the author's previous work (Knowles et al., 2020; Nikitas et al., 2020, 2021). This stratified selection process is also supported by best practice literature (Snyder, 2019; Van Wee & Banister, 2016). While the author acknowledges an element of subjective criteria in the material choice (i.e. selection bias is a natural human behavioural trend), this was a systematic and thorough process meaning to reduce bias and enhance the vibrancy, range and depth of the content.

AUTOMATED TRANSPORT'S PROJECTED IMPACTS ON THE DIFFERENT SPHERES OF SUSTAINABILITY

According to the agenda of the United Nations' Sustainable Development Goals, future urban planning with automated transport, which may evolve as the heart and soul of the smart city concept (Nikitas et al., 2021), must be underpinned by a new brand of sustainable thinking. This is because the current pace, nature and patterns of economic growth and the strategic policy-making tied to it are threatening our future and thus need to be reformulated so that they are bound by sustainability-enhancing preconditions.

The triple bottom line as introduced by Elkington (1998) is a theoretical and empirical benchmark that has been widely adopted and long acknowledged in both practice and research as a framework for recognising the three distinctive and complementary agendas that exist within sustainability (Svensson et al., 2018). These are namely the environmental, economic and societal pillars or spheres of sustainability (see Fig. 5.1).

A genuinely sustainable system only materialises in the space where the three spheres intersect, meaning that simultaneously all the environmental, economic and societal goals set are reached (Purvis et al., 2019). This chapter looks at some of the key themes that may be incorporated within the three spheres of sustainability namely: motor traffic, air pollution, energy consumption, employment,

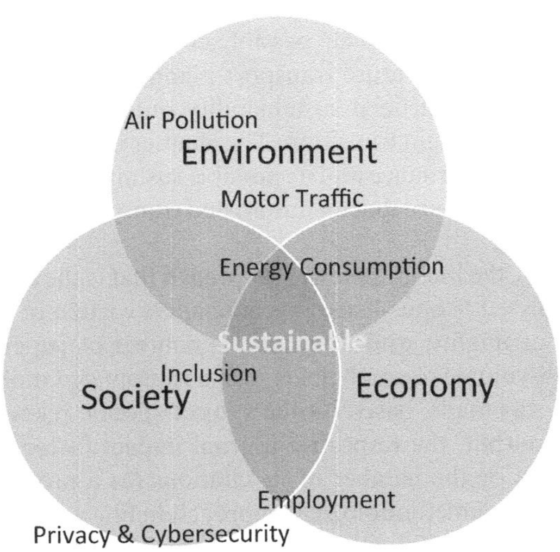

Fig. 5.1. Triple Bottom Line – The Three Spheres of Sustainability for Automated Transport.

accessibility, inclusion, cybersecurity and privacy. It should be noted that these agendas although classified for the means of this work under one of the three bottom line pillars, have dimensions and impacts that may extend to the other two pillars; thus their alignment in Fig. 5.1 places them in areas touching two or even three pillars of sustainability.

Environmental Sustainability

The environmental and energy-related aspects reflecting and affecting the potential introduction of automated transport in our built environment eco-systems do not allow us, for now at least, to reach a clear consensus on whether vehicle automation can yield benefits for the environment (Nikitas et al., 2021). However, the most likely scenario is that without a serious system-based shift to a primarily multimodal, public transit-based and alternatively fuelled transport paradigm the future might not look very promising in pro-environmental terms. CAVs, especially when working as a direct substitute for the unconnected, human-driven, conventionally fuelled, private automobile, are projected to impact the total *greenhouse gas (GHG) emissions* through their potential to:

a. Reduce vehicle ownership by popularising ride-sharing schemes if legislators, authorities and mobility providers make a regulated push to a 'sharing-based' direction;
b. Increase vehicle use intensity for four reasons: more people currently unable to drive or own them will have access to these vehicles; these vehicles will need no effort to 'be driven' and thus could allow people that otherwise would not want to put the physical and cognitive effort to drive use them for their trips; average trip costs in many simulated scenarios are projected to be smaller than current costs; regular public transport users may replace some of their public transport trips with car-centric ones in the scenario that CAVs are offered in shared schemes – this is a modal shift side-effect already picked up in Mobility-as-a-Service (MaaS) research (e.g. Alyavina et al., 2020). For a review of potential travel behaviour changes in CAVs, see Harb et al. (2021);
c. Create 'zombie car trips' where CAVs travel autonomously – and empty – around urban roads as they return to the depot or perform other tasks (Baron et al., 2022);
d. Extend dramatically urban sprawl, since according to Moore et al. (2020) there will be potentially up to a 68% increase in the horizontal spread of cities because of CAVs unless pro-active planning and policies are implemented to avert such consequence;
e. Increase total vehicle miles driven because of the reasons described in (b), (c), and (d), and the expected increased mobility needs of humanity because of the global population growth and the still rising urbanisation of the planet;
f. Impact parking needs and infrastructure: positively in case CAVs are shared and there are significantly fewer vehicles in the transport network that are always in motion meaning that urban space can be returned to other more liveable city functions; or negatively in case the number of vehicles rises

and more parking is needed or there is an increase in short-term parking requirements (parking briefly to wait for embarking or disembarking users in busy commercial or residential areas);

g. Change the vehicle fuel consumption rate by accelerating the transition to electrified transport – automation can be seen as the most powerful facilitator of electrification (Weiss et al., 2017) and actually this potential accelerating role of AVs in relation to the uptake of electrical vehicles (EVs) might be their single largest positive impact on the CO_2 emissions per km driven (Annema, 2020);
h. Eco-drive by helping to optimise braking, acceleration and deceleration and shift gears early;
i. Support platooning (i.e. road space and traffic flow optimisations) and improve routing performance although we are not there yet even for simpler interventions (Attard et al., 2016).

According to some optimistic projections based mostly on simulation studies since wide-spread experimentation of these vehicles per se is not possible yet, CAVs could support wider environmental goals. For instance, Pourrahmani et al. (2020) used a full-scale penetration of CAVs in the San Francisco Bay Area as a case study and estimated a 10% increase in car travel mileage (i.e. more people using a car more often for longer distances) and an 11% decrease in walk/bicycle trip mode share as well as significant emission reduction by several CAV-enabled mechanisms such as eco-driving [more than 30% carbon dioxide (CO_2) emission reduction], engine performance adjustment (over 20% CO_2 emission reduction), and platooning. Also aggressively programmed AVs that optimise road space requirements to the maximum may reduce GHG emissions by up to 26% from the current situation (Stogios et al., 2019), however, a more passive programming would mean an increase of 35% in emissions.

The most positive projections are always estimated when considering as the point of reference, a system that is based on Shared Autonomous Vehicles (SAVs) or even better on Shared Autonomous Electric Vehicles (SAEVs). SAVs could lessen to a significant degree GHG emissions by improving fuel efficiency, alleviating traffic congestion, facilitating the diffusion of alternative fuel vehicles, matching vehicle sizes to trip requirements, and reducing parking needs, among others (Chehri & Mouftah, 2019; Greenblatt & Saxena, 2015). SAEV fleets drawing power from city power grids may reduce GHG emissions by up to 73% and energy consumption by 58% compared to an automated fleet of Internal Combustion Engine Vehicles (ICEVs) and provide more inexpensive dollars-per-mile services (Bauer et al., 2018) supporting as many as 27 trips per day and 489 miles of daily average travel with very quick service response times (Loeb et al., 2018). However, even with SAEVs in place, the electricity consumed by the fleet and their market rollout would mean that they would have significant impacts on energy demand and, in turn, drive variation in energy cost and reliability, especially if the charging is unmanaged (Zhang & Chen, 2020). Thus, there is a need for research, development and investments in smart charging infrastructure and resource management (Nikitas et al., 2021).

On the other hand, though, CAVs could facilitate an increase in motor traffic by generating more vehicle miles travelled with all the negative consequences that this has on the environment with air pollution and also on public health and energy consumption. For example, using elasticities to simulate vehicle miles travelled and energy use impacts of full private CAV adoption under a range of possible changes to the fuel and time costs of travel, Taiebat et al. (2019) projected a 2–47% increase in travel demand for the average US household. Wang et al. (2018) modelled a case study scenario for the Greater Toronto and Hamilton Area to explore the environmental effects of AVs. The introduction of conventionally fuelled AVs resulted in higher vehicle miles travelled (3.6–5.4%) and GHG emissions (2.5%). However, in the case of an electrified fleet regional GHG emissions were reduced by 5%. It should be noted however that the latter results can only be meaningful if (a) the electricity powering these vehicles does not come from burning gas, coal or oil but from renewable energy sources (Kougias et al., 2020); and (b) the 'environmental costs' of batteries and their waste by-products are also considered.

Economic Sustainability

Economic sustainability is a multidimensional umbrella term that according to many scholars (e.g. de Oliveira Cavalcanti et al., 2017; Spangenberg, 2005) reflects elements such as economic growth, financial viability, profitability, cost minimisation and management, resource consumption and wealth accumulation. All of them are particularly important when considering the economic dimensions, implications and overall socio-economic architecture of the forthcoming CAV eco-system.

Urban traffic congestion is becoming an unmanageable problem with serious consequences including economic ones. Traffic delays and time spent in road congestion, in particular, are affiliated with economic and productivity losses (Weisbrod et al., 2003) that for the British economy alone were calculated to be around £8 billion a year. CAVs, which through their connectivity functions are able to talk to each other and to the other actors of their travel eco-system (Nikitas, 2021), may enable us to be more effective with the way traffic flow is managed. CAVs will be able to support platooning (i.e. decreasing the distances between vehicles and driving them together in groups) that would allow them to run on dedicated lanes and operate on a non-stop basis from origin to destination (Fernandes & Nunes, 2012). Improved routing that can get people from their origin to their destination efficiently (Zambrano-Martinez et al., 2019) and cooperative autonomous traffic organisation of the CAV fleets (Wang et al., 2020), where CAVs will work together as a single super-organism managed by a traffic controller, to coordinate and get collective benefits in support of a reduction in traffic delays (Nikitas et al., 2022) are powerful CAV qualities. CAVs, if implemented rightly, will also support the overall reduction of network traffic delays and peripheral costs like those attached to the public health system due to their traffic accident prevention capabilities that will reduce dramatically vehicle crashes, physical damages, traffic injuries and loss of lives (Bagloee et al., 2016).

AVs can actually serve as a significant employment disruptor, with profound economic consequences for the society. They will potentially remove millions of driving-related jobs, re-institute labour dynamics and make wider population reskilling, public engagement and awareness exercises an urgent and critically important task so that poverty and massive economic imbalance are not the vehicle automation's main side-effect (Nikitas et al., 2021). The list of the jobs in jeopardy is a lengthy one and includes, according to Nikitas et al. (2021): professional drivers in taxi, ride-sourcing, public transport and trucking industries and employees in vehicle manufacturing, rental, finance, retail and petrol stations, as well as road transport support industries. CAVs are forecasted to decrease expenses, and with that staff, on compulsory vehicle-related insurance policies (Fagnant & Kockelman, 2015), overhaul facilities, road traffic police forces, parking inspection provision, legal service providers (Davidson & Spinoulas, 2015), auto-repair industries and medical and emergency services (Clements & Kockelman, 2017). Nonetheless, new engineering, computing, data analytics and high-tech professional work opportunities could be generated by a transition to AVs that will counterbalance to some degree at least the losses of more traditional blue-collar posts.

If SAVs prevail as the primary business model of the driverless paradigm, car fleets will be significantly reduced (Narayanan et al., 2020) but will be able to provide lower fares to service users (Martinez et al., 2015) and more inexpensive dollars-per-mile trips (Bauer et al., 2018). These in purely economic and market terms, from the user perspective at least, could be a positive change that makes transport more affordable and inexpensive for all. AVs could also reduce commuters' value of travel time by 18–32% (8–14% for SAVs) compared to driving with changes being largest for suburban commuters, followed by their urban and rural counterparts (Zhong et al., 2020); this could lead to transport and land use planning implications that could promote urban sprawl.

Social Sustainability

Social or societal sustainability can be very difficult to quantify and monetise per se. Some studies and tools suggest that for calculating scenarios, researchers should primarily focus on environmental and economic sustainability (Zhong & Wu, 2015). Nevertheless, a qualitative coverage of some projected AV-related social impacts is helpful for drawing the bigger picture of the driverless transport transition and its sustainability aftermath.

The most important impact that automated mobility will have in this respect is the projected increase in the accessibility and social inclusion of mobility services. Increasing accessibility and balancing its distributional effects across space and social groups are two fundamental goals to make transport more sustainable and equitable (Dianin et al., 2021). More people currently unable to drive or maintain their own means of transport will potentially have access to CAV services (Nikitas et al., 2021). Elderly, children, people with physical disabilities and adults without a driver's licence, represent the largest groups of additional users (Meyer et al., 2017). However, people with a lack of digital skills may find it more difficult to use CAVs or avoid them altogether, so when CAVs become the norm, they may find themselves with less accessibility. Average trip costs in many simulated scenarios are projected to be smaller than the current costs of a car trip or relatively comparable with today's public transport fares making a CAV-based transport network more inclusive from an affordability standpoint.

Since CAVs require no effort 'to be driven', CAVs may have further impacts. Even within any shared provision format like that of a MaaS system, CAVs could allow people that otherwise would not want to put the physical and cognitive effort to drive, to use them for extra trips or for replacing more sustainable means of transport including public transport, walking and cycling (Alyavina et al., 2020, 2022). This will trigger the generation of more car trips and thus increase vehicle miles travelled. The social costs of increased emissions and noise (although a fleet of Electric AVs could be much quieter) that will be generated by these extra layers of inclusion are difficult to approximate (Wadud, 2017). This however adds another potential 'grey' dimension to the overall picture of the likely social sustainability impacts.

Rural communities are more dependent on transport services and systems than their urban peers and CAVs could ideally help address their needs. However, research that discusses CAVs' impact mostly focuses on urban transport with limited focus on rural areas since barriers concerning finance, transport infrastructure, policy and demographics may not allow CAVs to be as operational and impactful there (Prioleau et al., 2020). This could create social inequality for communities and individuals living in rural areas. This could be a factor for developing countries too, where the financial means would be much lower and the infrastructure to support a timely and efficient transition to CAVs would be too costly to build. This is a serious research gap in the existing driverless transport literature that with a few exceptions (e.g. Escandon-Barbosa et al., 2021; Sanaullah et al., 2017) fails to concentrate on developing countries and the Global South.

The transport workforce will experience first-hand the disruptive labour market aftermath of CAVs since these vehicles can make redundant driving and vehicle-related professions and create new layers of employability-related social exclusion (Nikitas et al., 2021). Recalibrating job accessibility via reskilling, awareness and training opportunities possibly targeting primarily blue-collar workers will be a way to pro-actively support a people-friendly transition to a driverless transport paradigm.

Finally, privacy (or its lack) and cybersecurity risks could be a set of complementary and severely unexplored societal AV-related concerns. The complex machine-led dynamics of CAVs will make them more prone to data exploitation and vulnerable to cyber-attacks than any of their predecessors, increasing the risks of privacy breaches and cyber security violations for their users (Liu et al., 2020). Privacy and cybersecurity considerations thus could be associated with fear, mistrust and undesirability of the AV concept and could create non-adoption or reluctance to use such services (Nikitas et al., 2022). Inter-institutional cooperation, awareness campaigns and trials for trust-building purposes and mandatory educational training and licencing for manufacturers and end-users could be a holistic solution that could mitigate these concerns (Liu et al., 2020).

On the other hand, an interconnected and automated transport system could be the platform enabling societies to maximise the potential of open big data in monitoring, controlling and planning for even better mobility services and supporting other mechanisms of the smart city as a whole.

DISCUSSION

The technological, behavioural and regulatory uncertainty around vehicle automation (Wadud et al., 2016) means that precise and accurate predictions of the likely impacts of automation on carbon emissions, transport energy consumption, job disruption, traffic delays and costs, accessibility, privacy and cybersecurity, are difficult to make. This chapter provides a snapshot of what is currently known and discusses some of the potential scenarios. It should be acknowledged though that every environmental, economic and social agenda item described herein can probably be significantly transformed by a currently unforeseen intervention or event. The COVID-19 pandemic that hit the world in 2020 and the unprecedented transport disruption that it caused (Kyriakidis et al., 2023) confirms this possibility. Similar threats arise from the increasingly unpredictable climate change events which are affecting various locations around the world every year, every month, every day. Indeed, it will not be a surprise if the main competitor of CAVs deployment in the future will be teleworking and a shift in residential locations and relocation of offices.

As the basis of an automated transport paradigm, CAVs have the potential, in theory, to facilitate unprecedented levels of eco-efficiency and function optimisation (e.g. eco-driving and platoon-based cooperation), radically reducing the transport sector's adverse energy and environmental impacts (Nikitas et al., 2021). However, depending on consumer choices (Harb et al., 2021; Morrow et al., 2014), the public transit, shared or private provision of CAV services (Etzioni et al., 2021; Taiebat et al., 2019), the adoption (or non-adoption) of electric, hydrogen and hybrid power sources to operate CAVs (Patella et al., 2019), and how these vehicles are programmed (Stogios et al., 2019) could instead result in increased motorised traffic. This will then lead to a net increase in energy consumption, traffic congestion, environmental pollution and their affiliated socio-economic costs. In all likelihood, the uptake of vehicle automation in isolation, and with an emphasis on automated car services, will result in moderate reductions in GHG emissions per mile that would be critically outweighed by a potentially high growth in vehicle miles travelled, compared to business as usual, through effects such as more car journeys and longer distances travelled, increased use of transport by underserved groups or empty-running vehicles, rising travel speeds, and lower value of time for commuting (Nikitas et al., 2021). This will subsequently lead to enhanced environmental, economic and societal burdens.

The response of the authorities that hold the responsibility for facilitating the transition to a driverless mobility paradigm to this negative but highly probable sustainability scenario is to ensure that vehicle automation is not a stand-alone, unfamiliar to many, intervention but one that is packaged with a few powerful counterparts of equal importance:

a. Connectivity that would enable AVs to communicate and cooperate (e.g. to facilitate functions such as platooning);
b. Alternative fuelling with an emphasis on electrification that makes automated transport oil-free and less polluting, provided that electricity is generated by renewable energy sources or by the utilisation of hydrogen fuel cell technology;
c. Multimodality that reflects a departure from a road transport-based monoculture looking at water, air and rail solutions; and
d. Sharing that should, in contradiction with a massive literature that is based on SAVs as shared automobile successors, be based on public transport – only mass public transport has the capacity to be genuinely sustainable in the long term so supportive policies to encourage public transport and reduce solo and empty car travel are needed.

CAVs in all their possible forms, may work much better in sustainability terms if they are packaged in public transit-dominant MaaS schemes that are actually designed to control and disincentivise when needed their car-based service offerings. These schemes need to be also complemented by other powerful mobility initiatives like Transit Oriented Development (TOD), travel demand management tools regulating parking and congested road networks, and traffic calming measures (Nikitas, 2021). Autonomous transport interventions, at the end of the day, need to be a key part of a holistic toolkit designed to safeguard and boost modal shift and help to break car-dominated travel behaviour habits; they need to be regulated in a way that allows them to be properly packaged and combined with sustainability-centric measures and not be an independent, stand-alone and uncontrollable mobility technology.

This means that besides the importance of technology development per se our transition-enabling efforts have to focus on policy reform and legislation, enhanced business models, and better education, training, information provision as mechanisms defining the principles and nature of transport automation (Nikitas et al., 2017, 2019, 2021; Thomopoulos & Nikitas, 2019). A change as big as that is multidimensional and needs a holistic approach and attention to detail even for agendas seemingly peripheral.

For instance, and in line with the recommendation for organically and compulsory integrating Shared Autonomous and Electrical Vehicle services in public transit-centred MaaS schemes, policy needs to design and create safeguard regulatory mechanisms to pro-actively control possible adverse side-effects like car-centric service dominance and vehicle mile increase. Bonuses and incentives should be given to sustainable road users, while 'excessive' conventional private car use equivalents should be limited, disincentivised or possibly banned when good public transit alternatives are in place. Key stakeholders, but also the general public (i.e. the potential automated transport users per se), need to become aware of the driverless transition and its challenges through education and training, information provision and consultation exercises. Care should be given to how these exercises should be designed to be inclusive, effective and appropriate to different populations.

This is because lifestyle factors, for instance, can play an important role in shaping AV attitudes, acceptance and potential usage; younger, urban residents who are more educated and technologically savvy are more likely to be early adopters of AV technologies and need a different engagement approach than older, suburban and rural individuals (Lavieri et al., 2017) who need more convincing methods 'to get on board in line with society's sustainability needs'. There are ways and channels however for all audiences to be informed, prepared and reskilled if necessary, to adapt to the changes that may be necessary so that challenges such as labour market disruption and lifestyle changes brought about by new driverless transport services can be addressed more effectively.

CONCLUSIONS AND FUTURE RESEARCH DIRECTIONS

Transitioning to a CAV-centric mobility paradigm, that has a diverse and severely under-explored spectrum of likely sustainability impacts and side-effects, is a lengthy, easy-to-go-wrong and complicated process. It requires a strategic and interdisciplinary approach centred around patience, resilience, determination, early action, and flexibility. Policy-makers, transport planners and mobility providers, not only in urban but also in rural environments, all share the responsibility for developing this approach. Vehicle automation is very much a story about driverless technology, but at the same time and in order to be successful, should be equally about nurturing an ethos of people-first policy-making and travel behaviour change. This is because, most likely, the adoption of vehicle automation alone, especially if it revolves around a travel behaviour regime that again is dominated by cars even if these are autonomous, will result in sustainability problems in all three spheres of the sustainability paradigm.

Autonomous, connected, alternatively fuelled, public transit-based mobility provided to well-informed, educated and change-prepared users, will need to be the norm of future travel service provision. This is the only way to achieve sustainability benefits such as traffic congestion decline, decreased atmospheric emissions, fewer contributions to global warming and climate change, optimised energy usage, greener fuelling, reduced resource overconsumption, reduced traffic delays, less accidents, minimisation of accessibility and employment inequalities and alleviation of privacy and cybersecurity issues.

It should be clear that CAVs, even if provided in the form of the nexus automation-connectivity-alternative fuelling-public transport to a well-informed user audience, may not necessarily be a panacea that will correct all that is wrong with transport today; as a matter of fact, it probably will not be able to fix alone some key sustainability problems that were described in this chapter and the book as a whole. Automated transport cannot and will not solve all the traffic problems that generate environmental, economic and social issues on its own. If, however, CAVs are used responsibly, consistently, wisely and with a real liveability-centric vision, they can be a significant tool, instead of a critical barrier to the transport toolkit for achieving a sustainable future transition.

More experimental studies at both the vehicle and the network system levels on the one hand and the user and non-user levels on the other, need to be produced before we can develop a better understanding of the nature, extent and significance of the sustainability effects that could be generated by autonomous transport systems. More specifically, understanding the particularities of AV technology implications for rural areas and the Global South, CAVs and their impact on urban sprawl and telecommuting, integration of CAVs in MaaS schemes and automated transport consequences in lifestyles, are all research gaps of critical importance and immediate policy interest.

REFERENCES

Alyavina, E., Nikitas, A., & Njoya, E. T. (2020). Mobility as a service and sustainable travel behaviour: A thematic analysis study. *Transportation Research Part F: Traffic Psychology and Behaviour*, *73*, 362–381.

Alyavina, E., Nikitas, A., & Njoya, E. T. (2022). Mobility as a service (MaaS): A thematic map of challenges and opportunities. *Research in Transportation Business & Management*, *43*, 100783.

Annema, J. A. (2020). Policy implications of the potential carbon dioxide (CO_2) emission and energy impacts of highly automated vehicles. *Advances in Transport Policy and Planning*, *5*, 149–162.

Attard, M., Haklay, M., & Capineri, C. (2016). The potential of volunteered geographic information (VGI) in future transport systems. *Urban Planning*, *1*(4), 6–19.

Bagloee, S. A., Tavana, M., Asadi, M., & Oliver, T. (2016). Autonomous vehicles: Challenges, opportunities, and future implications for transportation policies. *Journal of Modern Transportation*, *24*(4), 284–303.

Baron, O., Berman, O., & Nourinejad, M. (2022). Introducing autonomous vehicles: Adoption patterns and impacts on social welfare. *Manufacturing & Service Operations Management*, *24*(1), 352–369.

Bauer, G. S., Greenblatt, J. B., & Gerke, B. F. (2018). Cost, energy, and environmental impact of automated electric taxi fleets in Manhattan. *Environmental Science & Technology*, *52*(8), 4920–4928.

Chehri, A., & Mouftah, H. T. (2019). Autonomous vehicles in the sustainable cities, the beginning of a green adventure. *Sustainable Cities and Society*, *51*, 101751.

Clements, L. M., & Kockelman, K. M. (2017). Economic effects of automated vehicles. *Transportation Research Record*, *2606*(1), 106–114.

Davidson, P., & Spinoulas, A. (2015, July). Autonomous vehicles: What could this mean for the future of transport. In *Australian Institute of Traffic Planning and Management (AITPM) National Conference*, Brisbane, Queensland.

de Oliveira Cavalcanti, C., Limont, M., Dziedzic, M., & Fernandes, V. (2017). Sustainability of urban mobility projects in the Curitiba metropolitan region. *Land Use Policy*, *60*, 395–402.

Dianin, A., Ravazzoli, E., & Hauger, G. (2021). Implications of autonomous vehicles for accessibility and transport equity: A framework based on literature. *Sustainability*, *13*(8), 4448.

Elkington, J. (1998). Accounting for the triple bottom line. *Measuring Business Excellence*, *2*(3), 18–22.

Escandon-Barbosa, D., Salas-Paramo, J., Meneses-Franco, A. I., & Giraldo-Gonzalez, C. (2021). Adoption of new technologies in developing countries: The case of autonomous car between Vietnam and Colombia. *Technology in Society*, *66*, 101674.

Etzioni, S., Daziano, R., Ben-Elia, E., & Shiftan, Y. (2021). Preferences for shared automated vehicles: A hybrid latent class modeling approach. *Transportation Research Part C: Emerging Technologies*, *125*, 103013.

Fagnant, D. J., & Kockelman, K. (2015). Preparing a nation for autonomous vehicles: Opportunities, barriers and policy recommendations. *Transportation Research Part A: Policy and Practice*, *77*, 167–181.

Fernandes, P., & Nunes, U. (2012). Platooning with IVC-enabled autonomous vehicles: Strategies to mitigate communication delays, improve safety and traffic flow. *IEEE Transactions on Intelligent Transportation Systems*, *13*(1), 91–106.

Greenblatt, J. B., & Saxena, S. (2015). Autonomous taxis could greatly reduce greenhouse-gas emissions of US light-duty vehicles. *Nature Climate Change*, *5*(9), 860–863.

Harb, M., Stathopoulos, A., Shiftan, Y., & Walker, J. (2021). What do we (not) know about our future with automated vehicles. *Transportation Research Part C: Emerging Technologies*, *123*, 102948.

Knowles, R. D., Ferbrache, F., & Nikitas, A. (2020). Transport's historical, contemporary and future role in shaping urban development: Re-evaluating transit oriented development. *Cities*, *99*, 102607.

Kougias, I., Nikitas, A., Thiel, C., & Szabó, S. (2020). Clean energy and transport pathways for islands: A stakeholder analysis using Q method. *Transportation Research Part D: Transport and Environment*, *78*, 102180.

Kyriakidis, C., Chatziioannou, I., Iliadis, F., Nikitas, A., & Bakogiannis, E. (2023). Evaluating the public acceptance of sustainable mobility interventions responding to Covid-19: The case of the Great Walk of Athens and the importance of citizen engagement. *Cities*, *132*, 103966.

Lavieri, P. S., Garikapati, V. M., Bhat, C. R., Pendyala, R. M., Astroza, S., & Dias, F. F. (2017). Modeling individual preferences for ownership and sharing of autonomous vehicle technologies. *Transportation Research Record*, *2665*(1), 1–10.

Liu, N., Nikitas, A., & Parkinson, S. (2020). Exploring expert perceptions about the cyber security and privacy of connected and autonomous vehicles: A thematic analysis approach. *Transportation Research Part F: Traffic Psychology and Behaviour*, *75*, 66–86.

Loeb, B., Kockelman, K. M., & Liu, J. (2018). Shared autonomous electric vehicle (SAEV) operations across the Austin, Texas network with charging infrastructure decisions. *Transportation Research Part C: Emerging Technologies*, *89*, 222–233.

Martinez, L. M., Correia, G. H., & Viegas, J. M. (2015). An agent-based simulation model to assess the impacts of introducing a shared-taxi system: An application to Lisbon (Portugal). *Journal of Advanced Transportation*, *49*(3), 475–495.

Meyer, J., Becker, H., Bösch, P. M., & Axhausen, K. W. (2017). Autonomous vehicles: The next jump in accessibilities? *Research in Transportation Economics*, *62*, 80–91.

Moore, M. A., Lavieri, P. S., Dias, F. F., & Bhat, C. R. (2020). On investigating the potential effects of private autonomous vehicle use on home/work relocations and commute times. *Transportation Research Part C: Emerging Technologies*, *110*, 166–185.

Morrow, W. R., Greenblatt, J. B., Sturges, A., Saxena, S., Gopal, A., Millstein, D., Shah, N., & Gilmore, E. A. (2014). Key factors influencing autonomous vehicles' energy and environmental outcome. In G. Meyer & S. Beiker (Eds.), *Road vehicle automation* (pp. 127–135). Springer.

Narayanan, S., Chaniotakis, E., & Antoniou, C. (2020). Shared autonomous vehicle services: A comprehensive review. *Transportation Research Part C: Emerging Technologies*, *111*, 255–293.

Nikitas, A. (2018). Understanding bike-sharing acceptability and expected usage patterns in the context of a small city novel to the concept: A story of 'Greek Drama'. *Transportation Research Part F: Traffic Psychology and Behaviour*, *56*, 306–321.

Nikitas, A. (2021). Connected and autonomous vehicles: Priorities for policy and planning. In R. Vickerman & M. Attard (Eds.), *International Encyclopedia of transportation* (pp. 167–172). Elsevier.

Nikitas, A., Kougias, I., Alyavina, E., & Njoya Tchouamou, E. (2017). How can autonomous and connected vehicles, electromobility, BRT, hyperloop, shared use mobility and mobility-as-a-service shape transport futures for the context of smart cities? *Urban Science*, *1*(4), 36.

Nikitas, A., Michalakopoulou, K., Njoya, E. T., & Karampatzakis, D. (2020). Artificial intelligence, transport and the smart city: Definitions and dimensions of a new mobility era. *Sustainability*, *12*(7), 2789.

Nikitas, A., Njoya, E. T., & Dani, S. (2019). Examining the myths of connected and autonomous vehicles: Analysing the pathway to a driverless mobility paradigm. *International Journal of Automotive Technology and Management*, *19*(1/2), 10–10.

Nikitas, A., Parkinson, S., & Vallati, M. (2022). The deceitful connected and autonomous vehicle: Defining the concept, contextualising its dimensions and proposing mitigation policies. *Transport Policy*, *122*, 1–10.

Nikitas, A., Thomopoulos, N., & Milakis, D. (2021). The environmental and resource dimensions of automated transport: A nexus for enabling vehicle automation to support sustainable urban mobility. *Annual Review of Environment and Resources*, *46*, 167–192.

Nikitas, A., Tsigdinos, S., Karolemeas, C., Kourmpa, E., & Bakogiannis, E. (2021). Cycling in the era of COVID-19: Lessons learnt and best practice policy recommendations for a more bike-centric future. *Sustainability*, *13*(9), 4620.

Nikitas, A., Vitel, A. E., & Cotet, C. (2021). Autonomous vehicles and employment: An urban futures revolution or catastrophe? *Cities*, *114*, 103203.

Patella, S. M., Scrucca, F., Asdrubali, F., & Carrese, S. (2019). Carbon footprint of autonomous vehicles at the urban mobility system level: A traffic simulation-based approach. *Transportation Research Part D: Transport and Environment*, *74*, 189–200.

Pourrahmani, E., Jaller, M., Maizlish, N., & Rodier, C. (2020). Health impact assessment of connected and autonomous vehicles in San Francisco, Bay Area. *Transportation Research Record*, *2674*(10), 898–916.

Prioleau, D., Dames, P., Alikhademi, K., & Gilbert, J. E. (2020, November). Barriers to the adoption of autonomous vehicles in rural communities. In *2020 IEEE International Symposium on Technology and Society (ISTAS)* (pp. 91–98). IEEE.

Purvis, B., Mao, Y., & Robinson, D. (2019). Three pillars of sustainability: In search of conceptual origins. *Sustainability Science*, *14*(3), 681–695.

Sanaullah, I., Hussain, A., Chaudhry, A., Case, K., & Enoch, M. (2017). Autonomous vehicles in developing countries: A case study on user's view point in Pakistan. In N. A. Stanton, S. Landry, G. Di Bucchianico & A. Vallicelli (Eds.), *Advances in human aspects of transportation* (pp. 561–569). Springer.

Snyder, H. (2019). Literature review as a research methodology: An overview and guidelines. *Journal of Business Research*, *104*, 333–339.

Spangenberg, J. H. (2005). Economic sustainability of the economy: Concepts and indicators. *International Journal of Sustainable Development*, *8*(1–2), 47–64.

Stogios, C., Kasraian, D., Roorda, M. J., & Hatzopoulou, M. (2019). Simulating impacts of automated driving behavior and traffic conditions on vehicle emissions. *Transportation Research Part D: Transport and Environment*, *76*, 176–192.

Svensson, G., Ferro, C., Høgevold, N., Padin, C., Varela, J. C. S., & Sarstedt, M. (2018). Framing the triple bottom line approach: Direct and mediation effects between economic, social and environmental elements. *Journal of Cleaner Production*, *197*, 972–991.

Taiebat, M., Stolper, S., & Xu, M. (2019). Forecasting the impact of connected and automated vehicles on energy use: A microeconomic study of induced travel and energy rebound. *Applied Energy*, *247*, 297–308.

Thomopoulos, N., & Givoni, M. (2015). The autonomous car: A blessing or a curse for the future of low carbon mobility? An exploration of likely vs desirable outcomes. *European Journal of Futures Research*, *3*(1), 1–14.

Thomopoulos, N., & Nikitas, A. (2019). Smart urban mobility futures: Editorial for special issue. *International Journal of Automotive Technology and Management*, *19*(1–2), 1–9.

Toli, A. M., & Murtagh, N. (2020). The concept of sustainability in smart city definitions. *Frontiers in Built Environment*, *6*, 77.

Van Wee, B., & Banister, D. (2016). How to write a literature review paper? *Transport Reviews*, *36*(2), 278–288.

Wadud, Z. (2017). Fully automated vehicles: A cost of ownership analysis to inform early adoption. *Transportation Research Part A: Policy and Practice*, *101*, 163–176.

Wadud, Z., MacKenzie, D., & Leiby, P. (2016). Help or hindrance? The travel, energy and carbon impacts of highly automated vehicles. *Transportation Research Part A: Policy and Practice*, *86*, 1–18.

Wang, Y., Cai, P., & Lu, G. (2020). Cooperative autonomous traffic organization method for connected automated vehicles in multi-intersection road networks. *Transportation Research Part C: Emerging Technologies*, *111*, 458–476.

Wang, A., Stogios, C., Gai, Y., Vaughan, J., Ozonder, G., Lee, S., Daniel, P., Miller, E. J., & Hatzopoulou, M. (2018). Automated, electric, or both? Investigating the effects of transportation and technology scenarios on metropolitan greenhouse gas emissions. *Sustainable Cities and Society*, *40*, 524–533.

Weisbrod, G., Vary, D., & Treyz, G. (2003). Measuring economic costs of urban traffic congestion to business. *Transportation Research Record*, *1839*(1), 98–106.

Weiss, J., Hledik, R., Lueken, R., Lee, T., & Gorman, W. (2017). The electrification accelerator: Understanding the implications of autonomous vehicles for electric utilities. *The Electricity Journal*, *30*(10), 50–57.

Zambrano-Martinez, J. L., Calafate, C. T., Soler, D., Lemus-Zúñiga, L. G., Cano, J. C., Manzoni, P., & Gayraud, T. (2019). A centralized route-management solution for autonomous vehicles in urban areas. *Electronics*, *8*(7), 722.

Zhang, T. Z., & Chen, T. D. (2020). Smart charging management for shared autonomous electric vehicle fleets: A Puget Sound case study. *Transportation Research Part D: Transport and Environment*, *78*, 102184.

Zhong, H., Li, W., Burris, M. W., Talebpour, A., & Sinha, K. C. (2020). Will autonomous vehicles change auto commuters' value of travel time? *Transportation Research Part D: Transport and Environment*, *83*, 102303.

Zhong, Y., & Wu, P. (2015). Economic sustainability, environmental sustainability and constructability indicators related to concrete-and steel-projects. *Journal of Cleaner Production*, *108*, 748–756.

CHAPTER 6

REGULATORY FRAMEWORKS FOR TESTING AUTOMATED VEHICLES: COMPARATIVE ANALYSIS OF NATIONAL REGULATIONS AND KEY ASPECTS FOR A SUSTAINABLE IMPLEMENTATION

Dominik Schallauer[a], Aggelos Soteropoulos[a], Henriette Cornet[b], Wolfram Klar[a] and Alexander Fürdös[a]

[a]*Automated Mobility & Safety, AustriaTech – Federal Agency for Technological Measures, Vienna, Austria*
[b]*Connected and Cooperative Automated Mobility, UITP – International Association of Public Transport, Brussels, Belgium*

ABSTRACT

Many countries and regions have recognised the potential of automated transport as a solution to cover mobility needs in a sustainable way. They have implemented dedicated strategies and allowed trial operations of Automated Vehicles (AVs) within their national frameworks.

This chapter conducts an analysis of the legal frameworks for AV trial operations in 11 European countries. It reviews existing laws and regulations and

Sustainable Automated and Connected Transport
Transport and Sustainability, Volume 19, 101–117

ISSN: 2044-9941/doi:10.1108/S2044-994120240000019006

includes results from an online survey with national stakeholders and experts experienced in AV testing.

The results reveal very different approaches among European countries. Moreover, results indicate a stronger focus on technical safety aspects of the vehicles rather than on operational procedures and mobility integration, such as incorporating AV services into existing public transport systems.

This high level of disparity between the different European legal frameworks poses a considerable barrier to a rollout of the technologies and methodologies for AVs without cross-border and cross-supplier conflicts. Furthermore, when moving to the deployment of real services in the near future a common European framework and a stronger focus on operational procedures are essential for the implementation of automated transport services in order to cover the mobility needs of people in a more sustainable way (e.g. first/last mile to public transport).

European countries should further integrate operational aspects in the terms of services that are integrated in public transport, align deployment of AVs with national and local sustainability goals and focus on use cases beyond private vehicles to foster the transition to a more sustainable future of transport.

Keywords: Regulatory frameworks; automated vehicles (AVs); testing; sustainable implementation; European perspective; operational procedures

INTRODUCTION

Automated Vehicles (AVs) have the potential to not only increase road safety but also promote efficiency, inclusivity and on top of that have a smaller environmental impact (than current vehicles) (European Commission, 2021). However, before being able to achieve this, AVs need to be tested not only in the field but especially in real-life situations. Trial operations of automated transport with the requirement of a driver being in the vehicle (so-called 'safety driver') have already been enabled in many countries in Europe (Hansson, 2020) as well as in the United States, Australia (Lee & Hess, 2020) and several countries from Asia (Li et al., 2019). In the United States and Asia, even real-life operations of AVs without safety drivers on board are taking place (e.g. Waymo in Phoenix, GM Cruise in San Francisco or AutoX in Shanghai) (Loke & Rakotonirainy, 2021; Toensmeier, 2021) and are also planned in European countries like Germany (e.g. Mobileye in Munich) (Mobileye, 2021). To enable these tests and operations, countries as well as federal states had to implement regulations and specific legal frameworks (Lee & Hess, 2020).

In 2020, the UNECE made an effort to develop a harmonised regulatory framework for Automated Lane Keeping Systems (ALKS), the first binding international regulation that includes Level-3 functions (UNECE, 2021). The national procedures which allow the test and trial operation of AVs are still essential to test more advanced systems. The requirements and possibilities for testing and operation of AVs vary from country to country.

The enabling of testing and operating of automated transport and the development of respective rules for testing is mostly justified by the positive potentials of AVs to improve road safety but also accessibility for elderly and disabled people and those who are under-served by public transport (Lee & Hess, 2020). However, so far many studies stress the need for specific policies that accompany the development, testing and implementation of AVs to fulfil these potentials in order to attain a sustainable implementation of the technology (Pernestål et al., 2020).

Some studies (e.g. Easton, 2020; Hansson, 2020; Kouroutakis, 2020; Lee & Hess, 2020) already focus on differences in legislation for testing and operating AVs. Lee and Hess (2020), for example, investigated differences in legislation in Australia (Victoria), the United States (California), and Germany using a comparative analysis and found that all countries have a strong emphasis on the topic of safety management but that there are also some differences: Whereas for Victoria, Australia, and California, US safety management and functional safety plans for the testing of AVs are required, Germany does not require such safety plans. However, in Germany like in Victoria, Australia a human safety driver is required during tests, whereas California, USA only requires remote supervision of the AV with a continuous connection. For the safety management of AVs, California follows the US approach of self-certification of the broader safety requirements. In addition, the above-mentioned countries put an emphasis on regulations for data reporting and data maintenance with regard to the topic of safety (Lee & Hess, 2020). However, such an emphasis is not indicated for respective data or assessments with regard to a sustainable implementation of AVs for testing purposes.

Hansson (2020) focused on regulatory governance in the case of AVs in Sweden and Norway. She indicates that both countries have similar procedures for obtaining permission to test AVs, in which the applicants mainly have to prove that they meet a number of requirements, including traffic safety, vehicle control, and emergency procedures. While these requirements mainly focus on safety and security, for Norway it is reported that also other considerations will increase the probability for a permit, including amongst other things not competing with existing public transport routes when testing, ensuring that the application is in line with municipal strategies and creating a value for passengers (Hansson, 2020). These aspects align with the vision that AVs should be implemented as part of public transport in order to reach a sustainable implementation of automated transport with the maximum benefits in terms of efficiency, inclusiveness and impact on the environment (UITP, 2017).

Overall, existing studies show that a particular focus of regulations for testing AVs strongly lies on safety and security-related issues. Only in some countries, e.g. Norway, aspects with regard to sustainability are implicitly considered. However, a comprehensive analysis of national frameworks regarding test and trial operations of AVs especially from a European perspective is lacking. This chapter therefore, analyses national legal frameworks for the trial operation of automated transport for 11 European countries to give an overview of differences in the approaches and investigate to what extent sustainability aspects play a role. The main part of the analysis was conducted in 2020 and 2021, covering both passenger and logistics use cases.

METHOD

The analysis of national legal frameworks was based on a review of current laws, acts, ordinances and decrees that are enforced in the 11 selected European countries. The included countries are France, Germany, Austria, Sweden, Spain, The Netherlands, Czech Republic, Denmark, Finland, Greece, and Italy. For these countries, accompanying information, like government websites or codes of practices were analysed.

Besides this review, an online survey amongst experts on automated transport from the respective countries was performed. The goal was not only to review the theoretical requirements and possibilities, but also to include implicit rules and knowledge of experts who already gained experience with test and trial operations of automated transport services. The SHOW project, which is currently the most holistic project on shared automated mobility in Europe, offers a unique possibility to approach project partners from numerous European countries. The analysis presented in this chapter takes on certain aspects and results of the 'Analysis report on legal, regulatory, institutional frameworks' (SHOW, 2021).[1] For the 11 SHOW countries in which AVs will be deployed, the online survey was answered by 17 experts, working in the field of automated transport, based on their self-assessment of their local national frameworks.

The survey was carried out in August and September of 2020 and incorporated questions and assessments of the experts with regard to the topics of legislation and procedures for AV testing/operating, permit application process and requirements, vehicle technology, infrastructure and operational design domain (ODD), possibility of testing AVs in the context of logistics and platooning, human-centred aspects as well as data collection and recording. Since development in this field is very dynamic, the results of the survey are discussed in Section 4, also taking into account recent developments (after the survey was conducted) that target the real-world deployment of automated transport services.

The majority of questions in the online survey included closed-ended questions with various predefined answer options, with Likert scales often used for the answer options. In a few cases, the online survey also included open-ended questions, e.g. a description of the experts on the general reporting duties to the authorities in their country. For the analysis in this chapter, only results on specific aspects of the above-mentioned topics are presented.

Fig. 6.1 shows the number of participating experts by their professional background. It is noted that experts from various professional backgrounds participated in the online survey, however, the majority of experts (41%) work in research

[1]https://show-project.eu/wp-content/uploads/2021/04/SHOW-WP03-D-UIP-002-01_-_SHOW_D3.1_legal_regulatory_institutional_frameworks_SUBMITTED.pdf

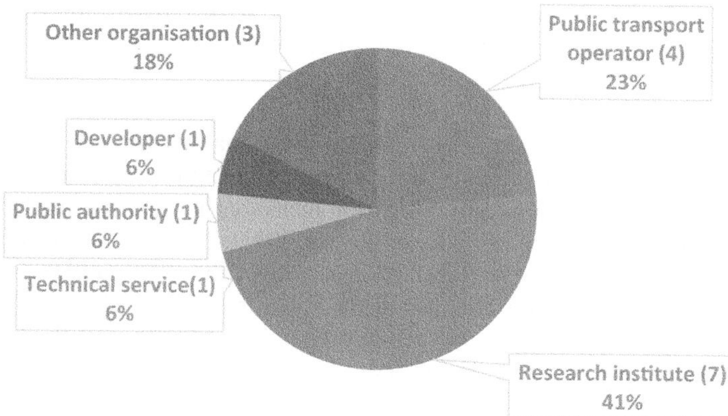

Fig. 6.1. Professional Background of the Participants of the Online Survey (Total Number and Share of Total Participants).

institutes and 23% work as public transport operators. In total, 17 experts participated, which included at least one expert for each country. For some countries, the assessments of two (Austria, Sweden, Spain, and Italy) or three (Finland) experts could be included.

RESULTS

Based on the review of the national frameworks in force, and experience from experts on automated transport in the respective countries, Section 3.1 summarises the current situation for test and trial operations in 11 European countries. In Section 3.2, the results of the online survey with national experts on automated transport are presented.

Review of National Frameworks in 11 European Countries

In the following section, the specific characteristics of the national frameworks of the 11 European countries, namely France, Germany, Austria, Sweden, Spain, The Netherlands, Czech Republic, Denmark, Finland, Greece, and Italy are described.

France

In France, the process to obtain approval for performing tests involves three ministries, i.e. the Ministry of Ecological Transition, the Ministry of Domestic Affairs and the Ministry of Economics. The process includes completing a comprehensive questionnaire (about 90 questions), where the applicant has to describe all aspects of the planned tests (République Française, 2018). Before the permit is issued the local road authority is also contacted. It is important to mention that the permit is only valid for the specified route and that the vehicle mode can always be tracked (automated or manual driving). The authorisation is granted for a maximum duration of two years. Another important aspect of the process is describing how the safety of the test is ensured (SHOW, 2021).

Germany

The process to obtain a testing permission in Germany involves an exemption according to §70 StVZO, which is only granted if technical and organisational documentation, risk analysis and mitigation measures have been provided by the applicant (German Federal Ministry of Justice, 2021). Additionally, the applicant has to define a data quality management system, which ensures an internal quality process, for example, for changes in the software. The applicant also has to provide documentation of an independent testing institution and the functions of the vehicles have to be demonstrated, also including simulated malfunctions of certain components (SHOW, 2021).

Austria

The Austrian legal framework 'AutomatFahrV' includes eight different test cases for test applications: 'Automated Minibus', 'AV for passenger transport', 'AV for the transport of goods', 'Motorway Pilot with Automatic Lane Changing', 'Motorway pilot with automated driving on slip roads', 'Automated Military Vehicle', 'Automated valet parking,' and 'Automated working machine' (BMK, 2022). Within the test cases, the specific requirements and possibilities for the respective test cases are defined. In Austria, the Ministry also invites experts to be part of a 'council of experts', who meet to discuss new test requests. It is important to note that requests that include aspects which go beyond the predefined test cases may involve a time-consuming process for amending the legal framework. Requirements for the test case 'Autonomous Minibus' include for example that the system has been tested for at least 1,000 test kilometres, that the speed is limited to 20 km/h, that passengers must be seated and that no commercial transport of passengers is allowed (SHOW, 2021).

Sweden

In Sweden, permits are only granted if the applicant can guarantee that traffic safety is ensured and that the trial operation does not cause any significant disruption or inconvenience to the surroundings (Transportstyrelsen, 2021). One specific characteristic of the Swedish legal framework is that the safety driver can also be located outside the vehicle. During the permit application process, applicants need to develop risk mitigation strategies. If the mitigation strategy involves handovers to a human driver, the applicant has to describe how this can be achieved whilst ensuring complete safety. In case of software changes, a new verification has to be carried out. Additionally, the transport agency also performs one-day site acceptance tests to check the system on-site before the permit is issued (SHOW, 2021).

Spain

The legal framework in Spain (INSTRUCTION 15/V-113) sets very specific requirements regarding the certification of test systems (DGT, 2019). This includes vehicle inspection, checks of the dynamic driving functionality, override of automated driving, lateral and longitudinal control or traffic sign recognition. The Spanish legal framework also explicitly mentions the mutual recognition with other EU Member states, if an equivalent prior control procedure has been used. The authorisation also must include the type of trial, route and all other conditions for the test. If the vehicle is driven outside the predefined route/zone, it must be in non-automated mode (SHOW, 2021).

The Netherlands

In the Netherlands, the legal framework – a law governing the experimental use of AVs ('experimental law on self-driving vehicles') imposes strict conditions, but also allows for tests where the driver is located outside of the vehicle (Dutch Government, 2021). The process involves working closely together with the vehicle authority, police, roads authority, and the Dutch Institute of Road Safety Research (RDW, 2021). Companies wishing to test AVs have to first convince and demonstrate that the test will be conducted safely (Dutch Government, 2021).

Czech Republic

So far, there is no dedicated legal framework for testing automated transport services including passengers. However, it is possible for car producers or technical services to apply for field-test permission, which is granted if the applicant has a homologation (approval of sale) certificate or confirmation that they have applied for homologation. To obtain the permission, the applicant also has to deploy a quality control system (CAD Knowledge Base, 2022a). This existing framework is mainly intended to be used for vehicle development by Original Equipment Manufacturers (OEMs) and not for tests including the transport of passengers with AVs. In 2020, the Ministry of Transport established an Ethics Committee that assesses the operation of AVs in the Czech Republic (Ministerstvodopravy, 2020).

Denmark

In Denmark, the permit application involves the Transport Agency, national police, director of public prosecution and road directorate and results in a final decision of the Danish parliament. During the process, a third-party assessor is also required. Furthermore, applicants need to consider how to inform other road users about the tests (Transport-ogBoligministeriet, 2017). Overall, the requirements and complexity for the applicant to fulfil all requirements are high, since many stakeholders and authorities are involved (SHOW, 2021).

Finland

Finland has a liberal legal framework regarding test applications of AVs. Each application is evaluated on an individual basis. The applicant has to give in-depth explanations on how traffic safety is ensured, including a general description of the trials, technical specifications of the vehicle and the road area (CAD Knowledge Base, 2022b). The application is handled by the Finnish Transport and Communications Agency (Traficom).

Greece

For previous research projects, Greek authorities established a legal framework which enabled tests in dedicated bus lanes. One specific characteristic of the Greek framework is that the tests need to be supervised by a public research or educational institute. The responsible ministry is the Greek Ministry of Transport, Infrastructure, and Networks, but besides the ministry, also the local municipality and the local road traffic police have to be involved (SHOW, 2021).

Italy

In Italy, the certification of new transport systems must include the description of technical standards and a risk assessment. The first step is to perform trials without passengers. If successful, a permit for transporting passengers is issued. Depending on the complexity of the system, the concrete procedure may vary from case to case (CAD Knowledge Base, 2022c). While the decree (DM70, 2018) mainly focuses on the Road Transport Infrastructures, a new decree (DM71, 2021) focuses on autonomous/intelligent vehicles (SHOW, 2021).

Overall, the review and analysis of the legal frameworks for trial operation of AVs in the 11 European countries show that usually the approval is granted on an individual basis and always includes different kinds of requirements. It also unveils that the main focus in all countries is to ensure the highest possible traffic safety. Nevertheless, some countries also include additional specific characteristics and national particularities. Latest information on testing regulations in the European countries can be found under the CAD Knowledge Base which also provides references to the respective online national repository of legal texts (CAD Knowledge Base, 2022d).

Online Survey with National Experts on Automated Transport

Following the review of national frameworks, the selected results of the online survey amongst national experts on automated transport are presented in this section. It should be noted that these are assessments from national experts, expressing their personal impressions and opinions on the different topics at the time when conducting the survey (August/September 2020).

BASIC ASPECTS REGARDING PERMIT APPLICATION PROCEDURES

Table 6.1 shows aspects regarding the general information and accessibility of information on test and trial permits. While information is available in most countries, so far there is no dedicated framework in the Czech Republic. When it comes to accessibility of the information, the results show a mixed picture. Easy access to the necessary information is not always guaranteed. For Germany and the Czech Republic, it is very difficult to find. Especially in the latter countries, there seems to be room for improvement for a more transparent presentation of the respective information. The third column shows how the experts rated the

Table 6.1. Basic Aspects Regarding Permit Application Procedures, $n = 17$ (Adapted from SHOW, 2021).

	Information Available	Accessibility of Information	Proactivity of Authorities	One-Stop-Shop or Several Involved?
France	Yes	+	++	One-stop-shop but several contacted and involved
Germany	Yes	0	++	Several
Austria	Yes	++	++	One
Sweden	Yes	+++	+++	Basically one, but e.g. also municipality could be involved
Spain	Yes	+++	+++	One authority and one technical service
The Netherlands	Yes	++	++	One
Czech Republic	No	0	0	Several
Denmark	Yes	++	++	Several
Finland	Yes	++	+++	One
Greece	Yes	++	+++	One
Italy	Yes	Opposing answers	Opposing answers	Opposing answers

Notes: Accessibility of information from 0 (very difficult to find) to +++ (very easy to find).
Proactivity of authorities from 0 (not at all) to +++ (totally).
Opposing answers = experts from one country gave opposing answers on a binary scale or completely opposing ratings on a Likert scale.

proactivity of national authorities – which means, how engaged and supportive they are perceived. In this regard, authorities in Sweden, Spain, Finland, and Greece have especially high assessments. In addition, the number of involved entities for the permit application varied between countries from one entity (one stop-shop) in Austria, the Netherlands, Finland or Greece to several involved entities in Denmark, the Czech Republic or Germany.

NUMBER AND TYPE OF REQUIREMENTS FOR PERMIT APPLICATION

Since applying for a permit goes hand in hand with fulfilling certain requirements, all experts were asked about their impression on the number of requirements that have to be fulfilled. The requirements were separated in different aspects, and Fig. 6.2 shows a comparison of four of those aspects, namely infrastructure-related, vehicle related, system control and data collection, and reporting requirements, for each country. In some countries, the number of requirements was rated very high for all four aspects, while other countries show more variety and a focus on individual aspects. For example, it can be observed that the number of vehicle-related requirements is rated generally high in all countries, but especially high in France, Germany, and Denmark. In addition, infrastructure-related requirements, that might include adaptations of physical or digital infrastructure, seem to play a very different role in the 11 countries, whereas the number of infrastructure-related requirements is rated high by the

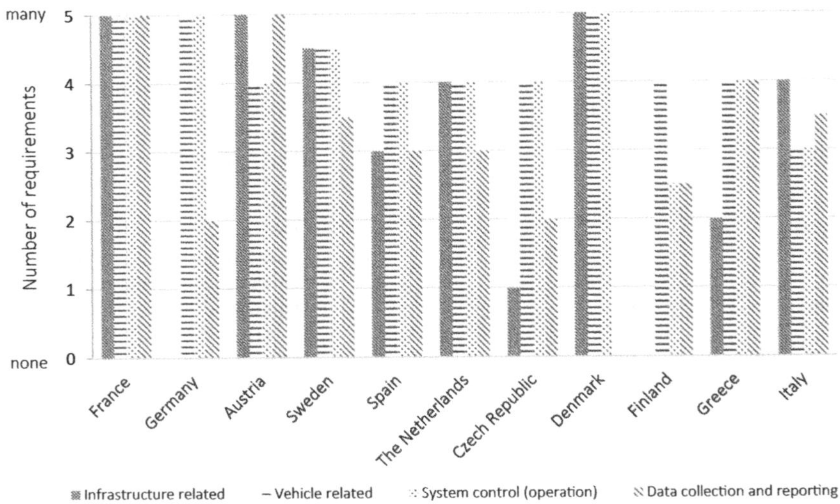

Fig. 6.2. Estimated Number of Different Types of Requirements for Permit Application, $n = 17$.

experts in countries like France, Austria and Denmark, it is rated particularly low in Greece and the Czech Republic. For applicants, it is important to not only consider conventional infrastructure elements like road markings, signage or traffic lights, but also digital infrastructure like C-ITS or 5G connectivity, which may require obtaining additional permits (SHOW, 2021).

Generally, the results show that all current national frameworks focus on different kinds of requirements. In addition to the categories provided in the questionnaire, experts in France and Spain mentioned additional requirements that have to be fulfilled. In France, there is a high number of requirements for IT systems and cybersecurity, regarding safety, risks and risk mitigation strategy, regarding the transport of passengers, routes, local authorisations and adaption of roadway and infrastructure. For Spain, additional requirements regarding functional safety, specific insurance and technical approval were mentioned.

POSSIBILITY OF TESTING AVs IN THE CONTEXT OF LOGISTICS

Table 6.2 shows the assessments of the national experts regarding the possibilities of testing logistics-related use cases. While transport of goods is possible in principle in most countries, some of the experts mentioned that there has been no operation in this context so far. Testing new vehicle concepts, where goods and passengers are transported at the same time is not possible in several countries, for example, in Greece and the Czech Republic, while this is possible in

Table 6.2. Aspects Regarding the Transport of Goods, $n = 17$ (Adapted from SHOW, 2021).

	Transport of Goods Allowed?	Transport of Passengers and Goods at the Same Time in One Vehicle	Platooning Allowed?
France	Yes	unsure	Yes
Germany	Yes	Yes	Yes
Austria	Opposing answers	Opposing answers	Opposing answers
Sweden	Yes	Yes	Yes
Spain	Yes	Opposing answers	Yes
The Netherlands	Unsure	Unsure	Yes
Czech Republic	Yes	No	No
Denmark	Yes	Yes	Yes
Finland	Yes	Yes	Opposing answers
Greece	No	No	No
Italy	Opposing answers	Opposing answers	Yes

Notes: Opposing answers = experts from one country gave opposing answers on a binary scale.

countries like Germany or Sweden. In addition, also platooning is not allowed in most countries, with only a few countries including France, Germany, Spain, and Greece allowing for the testing of platooning.

OPERATOR AND REMOTE OPERATION

Since remote operation plays an important role in the transition to real-life operation of automated transport services, Table 6.3 gives an overview of the current possibilities for testing. It becomes observable that in more than half of the countries, an operator must always be on board or physically connected to the vehicle. In this context, 'physically connected' includes specific use cases, where the operator may walk beside the vehicle. On the other hand, wireless remote operation

Table 6.3. Operator and Remote Operation, $n = 17$ (Adapted from SHOW, 2021).

	Operator Must Be on Board or Physically Connected	Wireless Remote Operation from Remote Locations Allowed
France	Yes	Yes
Germany	Yes	No
Austria	Yes	No
Sweden	No	Yes
Spain	Opposing answers	Yes
The Netherlands	Yes	Yes
Czech Republic	Yes	Yes
Denmark	No	Yes
Finland	Opposing answers	Opposing answers
Greece	No	Yes
Italy	Yes	Yes

Notes: Opposing answers = experts from one country gave opposing answers on a binary scale.

from a remote location is in principle allowed in the majority of countries (except Germany and Austria). However, this does not automatically allow the absence of an operator on board the vehicle.

REPORTING DUTIES

Table 6.4 gives an overview of the existence of reporting duties for the different countries. It is noticeable that reporting duties are in place in several countries, while this is not the case for Spain, the Czech Republic and Denmark. In countries where reporting duties are in place, usually incidents and accidents have to be reported, but there are also countries where reporting duties go beyond: For example, in Sweden this includes the evaluation of the trial after it is completed and in France, KPIs like km travelled or weather conditions have to be reported regularly. Overall, it seems that most of the countries have recognised that it is important to make use of the lessons learned during testing to further improving the legal framework.

DISCUSSION AND CONCLUSION

This chapter investigated the approaches in national legal frameworks for testing automated transport in 11 European countries by analysing the national legal frameworks and conducting an online survey with national stakeholders in these countries to give an overview of the differences in the approaches and investigate to what extent sustainability aspects play a role. Overall, the results of the analysis show very different approaches in the respective countries. This was especially revealed for the number of involved entities for the permit application varying from one entity (one stop-shop) in countries like Austria or the Netherlands to several involved entities in countries like Denmark or Germany as well as for the assessment on the number of requirements with regard to the vehicle, infrastructure, operation and data collection. Here, especially the stakeholders from

Table 6.4. Reporting Duties, $n = 17$ (Adapted from SHOW, 2021).

	Mandatory Regular Reporting
France	Yes
Germany	Undefined yet
Austria	Yes
Sweden	Yes
Spain	No
The Netherlands	Yes
Czech Republic	No
Denmark	No
Finland	Yes
Greece	Yes
Italy	Opposing answers

Notes: Opposing answers = experts from one country gave opposing answers on a binary scale.

France, Germany, and Denmark assessed the number of existing requirements for the vehicle as very high, while the respective assessment with regard to the number of existing requirements for the vehicle by national stakeholders for Italy was comparably low.

This shows a high level of disparity between the different European legal frameworks and their processes. The focus on data collection and the General Data Protection Regulation (GDPR) was also analysed by Costantini et al. (2020). However, harmonisation of regulations is required to ensure that technologies and methods developed for AVs can be rolled out in Europe without cross-border and cross-supplier conflicts. Thus, working towards interoperability of new technologies for AVs will lay a foundation for involvement with international standardisation organisations, which will further enhance European competitiveness (SRIA, 2021).

Fortunately, discussions and developments regarding harmonisation are progressing. In its session in early 2021, the UNECE Inland Transport Committee decided to mandate a Group of Experts to draft a new legal instrument on the use of AVs in traffic as a new legal instrument expected to complement the 1949 and 1968 Conventions on Road Traffic. In addition to the typical sections on definitions and final clauses, it will include a set of legal provisions for the safe deployment of AVs in international traffic (UNECE, 2022). The Group of Experts were given a two-year duration (with a possibility of extension) starting on 1 July 2021. Upon the completion of its term, this group will submit the complete draft of the new legal instrument to its supervising body, the Global Forum for Road Traffic Safety (WP.1) for consideration and decision. In regard to the approval of the amendments to the Vienna Convention, the integration of this text into the two Conventions of Vienna and Geneva should allow for a worldwide harmonisation of the conditions of circulation of autonomous vehicles by 2025 (French National Assembly, 2021).

Again, the discussions towards harmonisation target road safety, in particular the safety of vulnerable road users (that is also addressed in studies like Penmetsa et al. (2019) and Tabone et al. (2021) rather than addressing operational procedures or mobility aspects. Only for some countries, efforts towards the latter aspects are indicated. For example, for a few countries, results of the questionnaire with national stakeholders show that the testing of AVs with a transport of passengers and goods at the same time (e.g. Germany, Sweden, Denmark, or Finland) is possible. This also indicates that new vehicle concepts with AVs (beyond private vehicles) that integrate passenger and goods traffic (see, e.g. Ainsalu et al., 2018), which could potentially reduce traffic in the future, are possible to test. However, the analysis on the national frameworks in the countries reveals that these elements are scarce and that specific requirements are mostly focused on safety aspects of testing, rather than how AV services are operated, for example, if they are integrated into the current transport system or in combination with existing public transport services.

Recent developments, such as the entry into force of Implementing Regulation (EU) 2022/1426 for the type-approval of the automated driving systems (ADS) of fully AVs clearly highlight a progressive shift from testing to real deployments in certain use cases (European Commission, 2022). At some point, the first fully AVs for specific operational conditions will be proven safe enough for market introduction through this type-approval framework. For European Member States it will be very important to introduce the right frameworks for deployment and operation, since the regulation also states, that 'this regulation is without prejudice to the right of Member States to regulate the circulation and the safety of operation of fully automated vehicles in traffic and the safety of operation of those vehicles in local transport services' (European Commission, 2022). Besides ensuring the safety of operation, Member States should also consider sustainability aspects when integrating those vehicles into local transport services.

It is to be expected that legislations which integrate operational aspects in terms of services in their frameworks would foster countries and cities to explore and invest in sustainable modes of transport. For instance, authorisation could be facilitated for shared AV services that are integrated into the public transport system and thus lead to better accessibility and reduction of congestion and harmful emissions (UITP, 2021). Especially early implementation of AV operations in this context is key for facilitating broader service integration including booking, ticketing and route planning. In addition, legislation processes could include criteria for sustainable deployment aligned with their national and local sustainability goals.

Another aspect of sustainable mobility refers to the economic viability of AV services for cities, public transport authorities and operators. The results of the survey indicate that, in most countries so far, operators have to be on board of the vehicle, which could, especially in the early stages, increase the acceptance of using such vehicles (Kyriakidis et al., 2020). However, as long as the operators are obliged to deploy AV services with safety drivers, their market penetration will be low since the bus driver represents the largest part of operational costs (Ongel et al., 2019). In the latest legal developments in countries like Germany (BMVI, 2021) and France (République Française, 2021) as well as the new EU legal framework (European Commission, 2022), new legislations focus on allowing the operation of AVs without a driver in the vehicle, which is promising for future deployments.

Germany explicitly mentions specific use cases that should be fostered through this new regulation, like automated shuttle buses, people-movers, last-mile solutions or demand-responsive transport in off-peak times (BMDV, 2021). This indicates the increasing importance towards a more diverse implementation of AVs in tests and operations in Germany in the future. The focus lies clearly on testing and implementing use cases of shared AVs beyond private vehicles that could contribute to and support a more sustainable traffic in the future.

Gradually, these topics are also becoming increasingly important for other European countries and from the European perspective as a whole. The universal (legal) challenge is not only how to implement AVs safely in our transport systems, but also how to make sure that technology is really keeping its promises of being safer, eco-efficient and accessible to everyone (Roubik, 2018).

In summary, with regard to testing it is clear that current legislations especially, focus on safety. Here it will be important to increase testing in the future in order to achieve more public trust in the technology and for European countries to further harmonise their regulations on testing of AVs. In this regard, the proposal for an EU regulation on Artificial Intelligence (AI) could play a role in the near future. It includes provisions with regard to AI regulatory sandboxes, i.e. controlled environments that facilitate the development, testing and validation of innovative AI systems for a limited time before their market launch as well as requirements for example, data processing. When the regulation becomes effective, European member states will be compelled to comply with these provisions which will also affect the national legislation for testing of AVs.

However, with the new EU legal framework (European Commission, 2022) which will allow the market introduction of type-approved motor vehicles with automated driving systems, the consideration of operational and sustainability aspects needs to be intensified. These aspects have to be considered particularly when member states, which are still responsible for traffic rules and transport licencing, introduce frameworks for deployment and operation at a national level. In addition, the implementation on a regional and local level will become more and more important. From a sustainability point of view, for the determination of operating areas for the small-scale automated transport services, regulation and stakeholders on the regional and local level should focus on the integration of such services in current mobility goals and public transport. Here, measures and requirements to optimise pooling and reduce empty running of such vehicles as well as the establishment of multimodal and accessible hubs should be implemented, in particular together with stakeholders on a regional and local level. Furthermore, requirements for data monitoring are needed not only from a safety perspective (which is already done today when testing), but also with regard to mobility issues to adjust and optimise the automated transport services based on the data gathered from real-life operations to contribute to a sustainable local mobility system. In addition, not only will regulations be needed, but also soft law measures like incentives and agreements, for example, to integrate the new services in the street space with measures on curbside management and parking, as well as areas for boarding and alighting. This will increase the possibilities of pooling and optimising the street space for sustainable transport modes like walking and cycling and linking these transport modes with such automated services. All this is of particular importance to align the deployment of AVs with national and local sustainability goals and promote the transition to a more sustainable transport system.

REFERENCES

Ainsalu, J., Arffman, V., Bellone, M., Ellner, M., Haapamäki, T., Haavisto, N., Josefson, E., Ismailogullari, A., Lee, B., Madland, O., Madžulis, R., Müür, J., Mäkinen, S., Nousiainen, V., Pilli-Sihvola, E., Rutanen, E., Sahala, S., Schønfeldt, B., Smolnicki, P. M., Soe, R.-M., Sääski, J., Szymańska, M., Vaskinn, I., & Åman, M. (2018). State of the art of automated buses. *Sustainability*, *10*(9), 3118.

BMDV – Federal Ministry for Digital and Transport. (2021). *Gesetz zum autonomen Fahren tritt in Kraft*. https://www.bmvi.de/SharedDocs/DE/Artikel/DG/gesetz-zum-autonomen-fahren.html

BMK – Austrian Federal Ministry for Climate Action. (2022). *Automatisiertes Fahren Verordnung*. https://www.ris.bka.gv.at/GeltendeFassung.wxe?Abfrage=Bundesnormen&Gesetzesnummer=20009740

BMVI – Federal Ministry of Transport and Digital Infrastructure. (2021). *Entwurf eines Gesetzes zur Änderung des Straßenverkehrsgesetzes und des Pflichtversicherungsgesetzes – Gesetz zum autonomen Fahren*. https://www.bmvi.de/SharedDocs/DE/Anlage/Gesetze/Gesetze-19/gesetz-aenderung-strassenverkehrsgesetz-pflichtversicherungsgesetz-autonomes-fahren.pdf?__blob=publicationFile

CAD Knowledge Base. (2022a). *Regulations and Policies, Czech Republic*. https://www.connectedautomateddriving.eu/regulation-and-policies/national-level/eu/czech-republic/

CAD Knowledge Base. (2022b). *Regulations and Policies, Finland*. https://www.connectedautomated-driving.eu/regulation-and-policies/national-level/eu/finland/

CAD Knowledge Base. (2022c). *Regulations and Policies, Italy*. https://www.connectedautomateddriving.eu/regulation-and-policies/national-level/eu/italy/

CAD Knowledge Base. (2022d). *National level*. https://www.connectedautomateddriving.eu/regulation-and-policies/national-level/

Costantini, F., Thomopoulos, N., Steibel, F., Curl, A., Lugano, G., & Kováčiková, T. (2020). Autonomous vehicles in a GDPR era: An international comparison. In D. Milakis, N. Thomopoulos, B. van Wee (Eds.), *Advances in transport policy and planning* (Vol. 5, pp. 191–213). Academic Press.

DGT – Dirección General de Tráfico. (2019). *Instruction 15/V-113: Authorization to conduct tests or research trials of automated vehicles on roads open to general traffic*. http://catalonialivinglab.com/wp-content/uploads/2019/06/15v-113-vehiculos-conduccion-automatizada-english.pdf

Dutch Government. (2021). *Self-driving vehicles*. https://www.government.nl/topics/mobility-public-transport-and-road-safety/self-driving-vehicles

Easton, C. (2020). Autonomous vehicles: An analysis of the regulatory and legal landscape. In L. Edward, S. Burkhard, & E. Harbinja (Eds.), *Future law emerging technology, regulation and ethics* (pp. 313–340). Edinburgh University Press.

European Commission. (2021). *Strategic research and innovation agenda (SRIA)*. https://knowledge-base.connectedautomateddriving.eu/wp-content/uploads/2021/04/20210211-CCAM-Partnership-SRIA-v1.1.pdf

European Commission. (2022). *Commission Implementing Regulation (EU) 2022/1426 of 5 August 2022 laying down rules for the application of Regulation (EU) 2019/2144 of the European Parliament and of the Council as regards uniform procedures and technical specifications for the type-approval of the automated driving systems (ADS) of fully automated vehicles*. Retrieved January 24, 2023, from https://eur-lex.europa.eu/legal-content/EN/TXT/PDF/?uri=CELEX:32022R1426&from=EN

French National Assembly. (2021). *Le déploiement européen du véhicule autonome pour un renforcement des projets européens. Parliamentarian report of MP Damien Pichereau submitted to the French ministry of ecology and ministry of transport*. https://www.ecologie.gouv.fr/sites/default/files/rapport%20pichereau.pdf

German Federal Ministry of Justice. (2021). *Straßenverkehrs-Zulassungs-Ordnung*. http://www.gesetze-im-internet.de/stvzo_2012/

Hansson, L. (2020). Regulatory governance in emerging technologies: The case of autonomous vehicles in Sweden and Norway. *Research in Transportation Economics*, *83*, 100967.

Kouroutakis, A. E. (2020). *Autonomous vehicles: Regulatory challenges and the response from UK and Germany*. https://papers.ssrn.com/sol3/papers.cfm?abstract_id=3441264

Kyriakidis, M., Sodnik, J., Stojmenova, K., Elvarsson, A. B., Pronello, C., & Thomopoulos, N. (2020). The role of human operators in safety perception of AV deployment – Insights from a large European survey. *Sustainability*, *12*(21), 9166.

Lee, D., & Hess, D. J. (2020). Regulations for on-road testing of connected and automated vehicles: Assessing the potential for global safety harmonization. *Transportation Research Part A: Policy and Practice*, *136*, 85–98.

Li, S., Sui, P.-C., Xiao, J., Chahine, R. (2019). Policy formulation for highly automated vehicles: Emerging importance, research frontiers and insights. *Transportation Research Part A*, *124*, 573–586.

Loke, S. W., & Rakotonirainy, A. (2021). Automated vehicles, urban robots and drones: Three elements of the automated city. In S. W. Loke, & A. Rakotonirainy (Eds.), *The automated city* (pp. 69–108). Springer.

Ministerstvodopravy. (2020). *The operation of automated and autonomous vehicles will be dealt with by the newly established Ethics Committee.* https://www.mdcr.cz/Media/Media-a-tiskove-zpravy/Provozem-automatizovanych-a-autonomnich-vozidel-se?lang=en-GB

Mobileye. (2021). *Launching our Self-Driving Robotaxi in Germany.* https://www.mobileye.com/blog/self-driving-robotaxi-sixt-germany-iaa/

Ongel, A., Loewer, E., Roemer, F., Sethuraman, G., Chang, F., & Lienkamp, M. (2019). Economic assessment of autonomous electric microtransit vehicles. *Sustainability, 11*(3), 648.

Penmetsa, P., Adanu, E. K., Wood, D., Wang, T., & Jones, S. L. (2019). Perceptions and expectations of autonomous vehicles – A snapshot of vulnerable road user opinion. *Technological Forecasting and Social Change, 143*, 9–13.

Pernestål, A., Engholm, A., Kristoffersson, I., & Hammes, J. J. (2020). The impacts of automated vehicles on the transport system and how to create policies that target sustainable development goals. In *Shaping smart mobility futures: Governance and policy instruments in times of sustainability transitions.* Emerald Publishing Limited.

République Française. (2018). *Arrêté du 17 avril 2018 relatif à l'expérimentation de véhicules à délégation de conduite sur les voies publiques.* https://www.legifrance.gouv.fr/jorf/id/JORFTEXT000036868691/

République Française. (2021). *Décret n° 2021-873 du 29 juin 2021 portant application de l'ordonnance n° 2021-443 du 14 avril 2021 relative au régime de responsabilité pénale applicable en cas de circulation d'un véhicule à délégation de conduite et à ses conditions d'utilisation.* https://www.legifrance.gouv.fr/jorf/id/JORFTEXT000043729532

Roubik, M. (2018). *How Safe is safe enough? – Automated Mobility form a legal perspective. Beiträgezueinerökologisch und sozialverträglichen Verkehrsplanung 1/2018* (pp. 91–100). Research Center of Transport Planning and Traffic Engineering, Technical University Vienna. https://www.fvv.tuwien.ac.at/fileadmin/mediapool-verkehrsplanung/News/WCTRS/Publication/104_Roubik.pdf

SHOW. (2021). *D3.1: Analysis report on legal, regulatory, institutional frameworks. Deliverable of the Horizon 2020 SHOW project, Grant Agreement No. 875530.* https://show-project.eu/wp-content/uploads/2021/04/SHOW-WP03-D-UIP-002-01_-_SHOW_D3.1_legal_regulatory_institutional_frameworks_SUBMITTED.pdf

SRIA – CCAM Strategic Research and Innovation Agenda. (2021). *European leadership in safe and sustainable road transport through automation.* https://www.ccam.eu/wp-content/uploads/2022/01/CCAM-Partnership-SRIA-FINAL-2021.pdf

Tabone, W., de Winter, J., Ackermann, C., Bärgman, J., Baumann, M., Deb, S., Emmenegger, C., Habibovic, A., Hagenzieker, M., Hancook, P. A., Happee, R., Krems, D., Lee, J. D., Martens, M., Merat, N., Norman, D., Sheridan, T. B., & Stanton, N. A. (2021). Vulnerable road users and the coming wave of automated vehicles: Expert perspectives. *Transportation Research Interdisciplinary Perspectives, 9*, 100293.

Toensmeier, P. (2021, January). The coming driverless era. *Plastics Engineering.* https://doi.org/10.1002/peng.20430

Transport-ogBoligministeriet. (2017). *Lov om ændringaffærdselsloven.* https://www.retsinformation.dk/eli/lta/2017/696

Transportstyrelsen. (2021). *Automated Vehicles.* https://www.transportstyrelsen.se/en/road/Vehicles/self-driving-vehicles/

UITP – Union Internationale des Transports Publics. (2017). *Autonomous Vehicles: A potential game changer for urban mobility.* Policy Brief. https://cms.uitp.org/wp/wp-content/uploads/2020/06/Policy-Brief-Autonomous-Vehicles_2.4_LQ.pdf

UITP – Union Internationale des Transports Publics. (2021). *How to place public transport at the centre of the automated vehicle revolution.* SPACE Project Brief. https://cms.uitp.org/wp/wp-content/uploads/2021/10/Project-Brief-SPACE_OCT12021.pdf

UNECE – United Nations Economic Commission for Europe. (2021). *UN Regulation No. 157 – Automated Lane Keeping Systems (ALKS).* https://unece.org/transport/documents/2021/03/standards/un-regulation-no-157-automated-lane-keeping-systems-alks

UNECE – United Nations Economic Commission for Europe. (2022). *Group of Experts on drafting a new legal instrument on the use of automated vehicles in traffic.* https://unece.org/transport/road-traffic-safety/group-experts-drafting-new-legal-instrument-use-automated-vehicles

CHAPTER 7

INTEGRATED URBAN TRANSPORT PLANNING IN THE ERA OF AUTONOMOUS VEHICLES

Nikolaos Gavanas

Research Unit of Infrastructure, Technology Policy and Development, Department of Planning and Regional Development, School of Engineering, University of Thessaly, Volos, Greece

ABSTRACT

Apart from the challenges related to vehicle technology, the wide-scale deployment of autonomous vehicles (AVs) in cities is linked to unprecedented opportunities and unforeseen impacts. These refer to mobility conditions, infrastructure, land use, wider socio-economic factors, energy use and environmental and climate effects. AVs may affect all these in positive or negative ways, promoting or obstructing the promotion of specific aspects of sustainable urban development. An integrated planning framework is needed to maximise the positive impacts and mitigate the negative ones. The main obstacle in the process of developing such a framework is the absence of empirical data and experience from the implementation of this emerging technology. This chapter outlines the possible impacts of AVs and discusses their uncertainty and trade-offs in relation to sustainable urban development. The categorisation of impacts derives from the priorities of the UN Sustainable Development Goal (SDG) 11: Make cities and human settlements inclusive, safe, resilient, and sustainable. *The chapter also highlights the lack of data for the development of an evidence-based planning approach and suggests relevant recommendations to planners. In contrast to the current lack of data, the future abundance of Big Data collected by autonomous road transport systems is discussed in the context of future urban planning purposes. Based on the above, the chapter*

Sustainable Automated and Connected Transport
Transport and Sustainability, Volume 19, 119–137

ISSN: 2044-9941/doi:10.1108/S2044-994120240000019007

concludes by stressing the importance of an integrated urban transport planning approach that ensures a positive contribution of AVs to sustainable urban development. Hence, it offers valuable recommendations for policymakers in a range of fields.

Keywords: Autonomous vehicles (AVs); sustainable development; impact; data; transport planning; urban development

INTRODUCTION

The deployment of AVs in urban road networks has the potential of transforming mobility and accessibility conditions and is expected to impact the digitalisation, energy transition and decarbonisation of cities, affecting urban development both directly and indirectly (Milakis et al., 2020). As autonomous vehicle technology advances, many cities around the globe, mainly in the Western World (Canitez et al., 2018; van Acker et al., 2021), are testing autonomous transport systems in the framework of their smart city and sustainable mobility strategies, with San Francisco being one of the latest trials to be extended in 2023 for all-day operations. An indicative list of relevant trials which were conducted worldwide in the period 2016–2021 is presented in Table 7.1 to offer the relevant background for this chapter.

Despite the increasing number of trials in urban road networks, there is still great uncertainty concerning the anticipated impacts of the wide-scale deployment of driverless vehicles in cities (Thomopoulos & Givoni, 2015). The uncertainty is linked to the lack of real-life data, covering larger time periods and mixed traffic conditions (Gavanas, 2019). Unavoidably, a long time will be needed for the collection of the necessary data. Therefore, the effort to estimate the potential impacts of autonomous mobility on urban development may often lead to different, and sometimes controversial results since sustainable urban development usually operates in longer timeframes. For example, different assessments of the effect of AVs on Green House Gas (GHG) emissions range from an 80% reduction to a threefold increase under different scenarios (Greenblatt & Shaheen, 2015), demonstrating this extensive uncertainty.

Although there is no 'one-size-fits-all' approach, cities in the near future will have to find ways to integrate the potential of autonomous mobility into their paths towards sustainable urban mobility, while faced with the uncertainty of its overall impacts. In order to achieve that, urban transport planning needs to incorporate autonomous transport in the mid- and long-term (Fraedrich et al., 2019; Gridsted et al., 2022). It is worth mentioning that a Practitioner Briefing on the Sustainable Urban Mobility Plan (SUMP) Guidelines available through ELTIS[1] focuses on the difficulty to develop a long-term transport strategy that

[1]ELTIS is the Urban Mobility Observatory of the European Commission: https://www.eltis.org/

Table 7.1. Indicative List of Autonomous Driving Trials in Urban Areas in the Period 2016–2021 (*Source*: Compiled by Author from iMOVE Australia, 2022 and European Commission, 2022).

Year	Country	Project Name	Short Description
2021	Sweden	Gothenburg autonomous shuttle trial	Test of two autonomous shuttles along a pre-defined route.
	Denmark	Smart Bus in Aalborg	Pilot application of autonomous smart bus on a traffic-calming street.
2020	European Union (13 countries)	SHOW (Shared automation Operating models for Worldwide adoption)	Demonstration of integration of autonomous public transport, demand-responsive transport (DRT) and Mobility-as-a-Service (MaaS).
	UK	Project CAV Forth	Pilot application of an autonomous bus along a predefined route.
	Australia	Driverless Shuttle Trial: Newcastle NSW	Test of an autonomous shuttle along a predefined route.
2019	UK	Street Wise	Five autonomous passenger cars with security drivers for 24-hour service.
2018	European Union (4 countries)	AVENUE (AVs to Evolve to a New Urban Experience)	Test of AVs of various categories in urban road networks.
	Denmark	Denmark Technical University and Copenhagen Self-Driving Bus Trial	Test of autonomous bus inside a University campus and in an urban area.
	UK	AutopleX	Simulation and test of autonomous and automated vehicles along freeways and urban roads.
	UK	UK Autodrive	Test of autonomous passenger vehicles in controlled conditions along urban networks.
	Australia	Cohda Wireless driverless vehicle testing	Test of smart and connected driverless passenger vehicles.
	Canada	Autonomous	Research application of autonomous passenger vehicles to test under different weather conditions.
	USA	Drive.ai. On-Demand Self-Driving Car Service	Pilot programme for on-demand service provision by autonomous passenger vehicle.
2017	European Union (11 countries)	L3Pilot	Large-scale pilot of automated vehicles (SAE levels 3 and 4).
	Germany	Real-world trial of Automated Valet Parking	Test of AVs for parking in a controlled environment.
	Australia	La Trobe University Driverless Bus Trial	Test of autonomous mini-bus inside a University campus.
2016	European Union (6 countries)	CityMobil2	Test of autonomous mini-buses along dedicated routes.

includes cooperative, connected and automated mobility (CCAM). It also highlights the absence of empirical data and appropriate planning tools and models for the detailed assessment of CCAM impacts (Rupprecht Consult, 2019a). As mentioned in the overall SUMP guidelines (Rupprecht Consult, 2019b), planning for sustainable urban mobility should ensure consistency and complementarity with land use and urban planning. This integrated approach to planning stems from the well-studied interaction between the land use and transport systems (Rode et al., 2017), briefly described as follows: The organisation of the land use network implies a spatial separation between activities. The transport system provides access to these activities while changing the overall spatial accessibility conditions and, thus, creating opportunities for new connections and affecting land use location decisions (Wegener & Fürst, 1999). The possible impacts from the wide-scale deployment of AVs, as discussed in the following sections, are observed through the lens of the transport and land use interaction, as they touch upon the topics of mobility, the built and natural environment and spatial development.

In the European Union (EU), the current urban mobility policy framework, as described by (European Commission, 2021), makes an explicit link by referring to the effect of urban mobility on sustainable urban development and on the pursuit of the UN Sustainable Development Goal 11: *Make cities and human settlements inclusive, safe, resilient and sustainable* (United Nations, 2022). The framework also highlights the need for synergies between urban transport and spatial planning. Concerning AVs, the document mentions the key role of zero-emission, connected and automated mobility to support both the sustainable urban mobility objectives and the EU decarbonisation targets, and also outlines specific strategies for the promotion of CCAM.

In this context, the current chapter describes the possible wider impacts of AVs on sustainable urban development and discusses their uncertainty and the trade-offs. The chapter also highlights the lack of data and suggests alternative generic approaches to tackle the problem. The big data collected by the autonomous road transport system is discussed as a means to support future urban planning purposes. The chapter concludes by stressing the importance of an integrated urban transport planning framework that safeguards the potential contribution of AVs to sustainable urban development.

METHODOLOGICAL FRAMEWORK

The objective to provide planners with a comprehensive overview of the possible impacts of autonomous transport on sustainable urban development is based on a thematic literature review. In order to fully cover a range of these impacts in a systematic way, they are categorised according to the targets of UN SDG 11: 'Make cities and human settlements inclusive, safe, resilient and sustainable' (United Nations, 2022). These targets are directly linked to the urban transport system or to other aspects of urban development that may be affected by the implementation of driverless vehicles, as described in Table 7.2.

Table 7.2. Framework for the Relation of UN SDG 11 Targets to Autonomous Mobility (*Source*: Own Elaborations from United Nations, 2022).

Target	Short Description	Relevance to Autonomous Road Transport
11.2	By 2030, provide access to safe, affordable, accessible and sustainable transport systems for all, improving road safety, notably by expanding public transport, with special attention to the needs of those in vulnerable situations, women, children, persons with disabilities and older persons.	• Autonomous road transport aims to remove human error from driving and improve road safety for both private and collective transport. • Autonomous cars aim to be accessible to wider groups of travellers, such as people with mobility impairments or without a driving licence. • Autonomous collective transport may enhance the accessibility and demand-responsiveness of mobility services to vulnerable groups.
11.3	By 2030, enhance inclusive and sustainable urbanization and capacity for participatory, integrated and sustainable human settlement planning and management in all countries.	• Autonomous road transport is expected to change travel patterns and potentially impact location choices. Planning will be required to mitigate shifts from public transport and active travel to AVs. • As shared AVs need less surface for driving and parking operations and have the ability of remote parking, more space could become available in cities. • Participatory planning and stakeholder engagement are needed for the introduction of AVs in a way that would correspond to the mobility needs of people and businesses.
11.6	By 2030, reduce the adverse per capita environmental impact of cities, including by paying special attention to air quality and municipal and other waste management.	• Autonomous road transport, by changing the way that people access and use mobility, could affect the environmental footprint of urban transport. • Scientists claim that a sustainable autonomous transport future should be shared, connected and electric, based on renewables, in order to enhance energy efficiency and support decarbonisation.
11.7	By 2030, provide universal access to safe, inclusive and accessible, green and public spaces, in particular for women and children, older persons and persons with disabilities.	• The space that may be freed up due to autonomous road transport (see also Target 11.3) can be used to increase the availability of public space. • The ability of AVs to service people with reduced mobility could enhance their accessibility to green and public spaces. • Attention however will be required to avoid the potential increase in congestion due to increased car use.
11.a	Support positive economic, social and environmental links between urban, peri-urban and rural areas by strengthening national and regional development planning.	• Autonomous road transport may affect the conditions of spatial accessibility and the connectivity between urban, peri-urban and rural areas, provided it is appropriately integrated into strategic planning.

By using Table 7.2 as a framework, the analysis of possible impacts of autonomous transport on sustainable urban development comprises the following categories:

- Mobility conditions and road safety,
- Infrastructure and spatial development, and
- Energy, environment and climate change.

Apart from the above categories, the potential impacts of autonomous transport refer to a series of horizontal challenges of socio-economic sustainability, i.e. business and economic development, employment and skills, personal data and security, and ethics and liability. These impacts do not directly affect the integrated transport planning process but should be closely examined by the planner. The synthetic analysis of the categorised impacts and the concerns about data availability lead to the planning recommendations and the conclusions on the integration of autonomous mobility into a city's transport planning framework.

POSSIBLE IMPACTS OF AUTONOMOUS ROAD TRANSPORT ON SUSTAINABLE URBAN DEVELOPMENT

This section focuses on selected potential AV impacts as reported in academic literature, policy documents and industry reports.

Mobility Conditions and Road Safety

The new elements that the deployment of autonomous road transport is expected to bring to urban mobility are the following:

- Accessibility by more users in comparison to conventional cars,
- Engagement to other activities, instead of driving, while travelling,
- Uniform driving behaviour,
- Driverless and remote parking and/or refuelling/charging, and
- Removal of human error as a cause of road accidents.

Fully AVs (Automation Level 5) could be operated by people without a driving licence and, thus, can be accessed by a greater share of potential users compared to human-driven cars. This is estimated to be more efficient in the case of shared AVs (Simpson et al., 2019). AVs can be used to improve the mobility conditions of people with disabilities, younger or older travellers and, in general, people with reduced mobility, while they can act as enablers of Mobility-as-a-Service (MaaS) (Williams, 2021). On the other hand, the improved accessibility of autonomous cars may lead to the increase in the number of daily car trips with a potentially negative impact on congestion (Metz, 2018).

Autonomous cars are self-driven, allowing for all vehicle users to perhaps conduct other activities (social or recreational activities, work etc.), while on the move. This is expected to affect the value of time, especially for commuters, professional drivers and other professionals that use driving as part of their daily routine (e.g. service vehicle drivers, couriers, etc.) (Meyer et al., 2017). It may also lead to the increase in travel time, travel distance and trip frequency, because drivers will be able to take advantage of the time that they spend in traffic, despite the potential negative impact on the environment and health, of longer commutes or travel in general. In the case of autonomous shared cars and demand-responsive passenger and freight services, the vehicles will probably move for longer periods and cover larger areas in order to provide on-demand services, leading to the increase of motorised traffic in cities (Nikitas et al., 2021).

In contrast to the above, the deployment of connected and autonomous mobility is expected to lead to the normalisation of traffic flows with positive impacts on the reduction of congestion, due to the more stable driving behaviour and speed profiles of driverless vehicles (Fagnant & Kockelman, 2015). Moreover, drivers nowadays spend a significant amount of time cruising for parking at their destination. The search for parking spaces corresponds to 30% of urban road traffic in Europe, while parked cars occupy a significant part of roadway capacity (Transport & Environment, 2017). The ability of AVs to reach parking or refuelling/charging stations after the drop-off of their passengers may reduce parking-related congestion and free roadway capacity (Lu et al., 2020).

Regarding road safety, by removing human errors from driving, AVs are aiming to reduce road fatalities. According to a study in the US, the penetration of autonomous cars by 10%, 50% and 90% is expected to save 1,100, 9,600 and 21,700 lives per year respectively (European Commission, 2018a). However, AVs have also contributed in road fatalities, although in low numbers yet, and some estimates show that even in the best-case scenario, approximately 30 years would be needed to replace the current fleet of conventional vehicles with autonomous ones. During the transition period, vehicles of different levels of automation will co-exist and share the road infrastructure with active transport and micromobility modes (Martens et al., 2021). Researchers argue that during this period the impacts on comfort, safety and congestion may be negative because of the infrastructure requirements and the driving capabilities of different modes and technologies (Soteropoulos et al., 2019). For example, autonomous cars need less lane width and headway distance than conventional cars. This may have an adverse effect on the mobility conditions of both human-driven cars and active transport and micromobility users (Parkin et al., 2018).

Infrastructure and Spatial Development

Recent research indicates that, in order for AVs to operate safely and efficiently without compromising the safety and comfort of other road users, the traffic, parking and refuelling/charging physical and digital infrastructure should be appropriately adjusted (Bosch et al., 2018). For example, if an autonomous transport system aims to increase the accessibility of persons with reduced mobility, it is essential that the pedestrian infrastructure should also be adjusted to ensure seamless door-to-door access for all travellers. The same principle applies for multimodal transport hubs in the case that the AVs operate as part of an integrated multimodal transport system.

Apart from transport infrastructure, the debate on the promotion of connected and automated mobility in combination with electromobility stresses the need to integrate the urban transport, energy and telecommunication networks in the context of Smart City infrastructure (World Economic Forum, 2018). The integration with telecommunication networks is required for AVs to effectively exchange big data between themselves and with other infrastructure elements (e.g. traffic management and road safety infrastructure, real-time information systems, fleet management systems, etc.). It is also considered essential for electric AVs and their charging infrastructure to be connected to the city's power grid, allowing for demand-responsive charging and the provision of ancillary services back to the grid (Iacobucci et al., 2018). There are however still concerns about the ability of energy grids to support entire fleets of electric vehicles in parts of cities where density is already high or demands on the infrastructure is considerable and struggles with seasonality, especially in popular tourist destinations.

AVs could need less space to navigate in traffic or to conduct parking manoeuvres in comparison to human-driven vehicles (Nourinejad et al., 2018; Ruso et al., 2019). Furthermore, they can search for parking spaces at a distance from the traveller's destination, which may be located in a crowded, central area, a tourist area, a park, a health or educational facility and in any other area where planners aim to reduce traffic. Effective planning and management would be required to contain any increase in traffic (empty return trips) that these types of services might generate. Thus, the wide-scale deployment of AVs, mainly in the context of collective (public and/or shared) transport, could potentially free up public space, which can be used for the development of active transport and micromobility infrastructure or other land uses, e.g. urban green spaces and linear parks. This would depend on integrated transport and urban planning decisions taken. Another potential advantage of the potential ability of AVs to move and manoeuvre in narrower spaces is the service of roads of poor geometrical characteristics, such as some historical centres or densely constructed neighbourhoods. A balance between the benefits accrued from potentially increased accessibility and autonomous transport will be necessary to reduce any uncontrolled increase in motorised traffic, as this would eventually need additional roadway capacity. Moreover, the use of exclusive lanes for the safe operation of AVs should be carefully considered as it will affect the re-organisation of existing roadway capacity and use of public space (Anastasiadou, 2021; World Economic Forum, 2018).

As discussed in the previous section, autonomous mobility may enhance the accessibility, flexibility, safety and comfort conditions of daily urban mobility and, thus, increase the number and frequency of trips and the distance and duration that people are willing to travel. This is expected to affect the location choices of households and companies, influencing the spatial organisation of activities and, consequently, the urban form. Depending on the relation between the type of AV deployment (private, shared or public transport) and the city's specific characteristics and priorities, different urban development patterns may be observed, ranging from suburbanisation and land use zoning (similar to the impact from the private car dominance in western cities of the post-World War II period) to compact city development, mixed-land uses and the development of

15-minute cities (especially in the case of collective autonomous transport systems) (Hawkins & Habib, 2019; Heinrichs, 2016).

According to the review of scenario studies conducted by Stead and Vaddadi (2019), if car ownership remains dominant and AVs are introduced as private cars, this would lead to an increase in demand for more road and parking space, especially in central areas. On the other hand, the shift to shared autonomous mobility could lead to the reduction of space requirements and free up space for other developments. All this despite evidence suggesting that shared car use has not increased vehicle occupancy and might indeed decrease it even further with AVs circulating without passengers (e.g. returning to allocated parking location, or reaching the next customer). In both cases, planners should consider that the advantages of AVs for the user may affect the demand for active and conventional public transport, which is important for the service of mobility needs in compact cities (van der Waals, 2000). For the promotion of a more compact urban development pattern, the deployment of shared autonomous mobility should be complemented by policy interventions, including investments and incentives to promote MaaS, public and active transport and shared micromobility, and measures for the internalisation of external costs from transport and land use planning interventions.

Energy, Environment and Climate Change

In the late 20th century, the World Commission on Environment and Development set the principles for sustainable global development and the need to adjust growth policies in relation to the limited environmental resources and the ability of the biosphere to absorb the effects of human activities (World Commission on Environment and Development, 1986). It emphasised the challenges of increasing urbanisation and the need to examine new urban transport policies. In 2015, the United Nations agreed on the 2030 Agenda for Sustainable Development and publishing the 17 Sustainable Development Goals (SDGs) aimed at international cooperation and country-wide actions to achieve substantial and tangible results in socioeconomic and environmental sustainability (United Nations General Assembly, 2015). In the same year, the Paris Agreement was adopted at the United Nations Climate Change Conference (COP21), an international treaty to face climate change and limit global warming (United Nations, 2015). These strategies are affecting urban development and mobility policies at the global level. For example, among the SDGs, the aforementioned Goal 11.

In this context, it is essential to consider the possible effects of the deployment of autonomous mobility on the environmental sustainability of cities and their role in global warming and energy transition. For example, the Chinese Ministry of Industry and Information Technology aims for the car industry to develop 2025, smart and connected vehicles with the capability to cut energy consumption by more than 10% and GHG emissions by more than 20% (ERTRAC Working Group: 'Connectivity and Automated Driving', 2019). In Europe, the EU Smart and Sustainable Mobility framework aims by 2030 for at least 30 million zero-emission vehicles to operate on the European road network as well as for the wide-scale deployment of automated mobility, while by 2050 almost all road vehicles will produce zero emissions (European Commission, 2020). It will be a challenge to ensure clean, renewable energy to support this shift. It should be noted that the total energy efficiency and environmental effect of electric vehicles, compared to vehicles with internal combustion engines, depends on the way that the electric energy is produced, as well as the integration of the vehicle charging infrastructure into the city's electricity grid and the effective management of energy resources.

There is a potential for automation to increase congestion during the transition period, when vehicles of diverse automation levels co-exist in the road network, due primarily to the different driving capabilities and requirements of the vehicles and their drivers. Thus, fuel consumption and emissions may increase locally. Moreover, if the volume, duration, length and frequency of trips increase due to the enhanced accessibility, safety and comfort conditions provided by AVs, the total impact on energy consumption and the environment may be negative (Wadud et al., 2016). This increase in vehicle kilometres is also linked to a shorter life span of each vehicle, especially in the case of shared AVs that will be used more intensively than private ones. Therefore, the vehicle's environmental footprint from a life-cycle perspective depends on the sustainability of the manufacturing and maintenance processes, and the potential reusability of materials.

Horizontal Challenges of Socio-Economic Sustainability

The socio-economic sustainability challenges do not influence the transport planning process in a direct way. Nonetheless, they should be considered in the process of engaging the stakeholders to be involved in a participatory planning procedure (Kaplan et al., 2023). As an indicative example, Sustainable Urban Mobility Plans aim *to satisfy the mobility needs of people and businesses in cities and their surroundings for a better quality of life* (Rupprecht Consult, 2019b). Contrary to the traditional transport planning process, conducted solely by expert practitioners, the SUMP process includes citizen involvement and stakeholder engagement from an early stage to learn from local knowledge and experience, and to incorporate society's needs and expectations in the design of solutions, as well as to ensure public acceptance, adoption, and support of the planned interventions. However, both aspects of public participation in planning for autonomous mobility are faced with the lack of experience from the daily use of AVs in cities, and the uncertainty of possible impacts as highlighted in Table 7.3. In order to cope with these challenges, the deployment of autonomous mobility should involve the exchange of experiences between cities and the engagement of a broad range of stakeholders, including local authorities, planners, transport service providers and professionals, vehicle and equipment manufacturers and service providers, technology companies, insurance and legal consultants, retail companies and representatives from the civil society (Rupprecht Consult, 2019b). The systematic collection of data regarding citizen and business perceptions and the provision of information through widely accessible channels can increase the readiness of both planners and the public (van Acker et al., 2021).

Table 7.3. Main Challenges of Autonomous Mobility for Socio-Economic Sustainability (*Source*: Own Elaboration).

Aspects of Socio-Economic Sustainability	Challenges Related to Autonomous Mobility
Business and economic development	The deployment of connected and autonomous mobility incorporates new synergies between the automobile and the telecommunication sectors and new business models for service providers in order to capitalise on the technological evolution in the Internet of Things (IoT), the industry digitalisation, the 5G implementation and big data (European Commission, 2016). The implementation of AVs is expected to introduce changes in the business models of the insurance sector due to the shifting of road accident liability from the driver to the car's manufacturer, technology provider and software developer, as further discussed below.
Employment and skills	AVs are fully self-driven and their wide-scale deployment is expected to lead to significant job losses for professional drivers, while upskilling opportunities are possible with the appropriate policy interventions (International Transport Forum, 2017). Changes in job characteristics of the car manufacturing industry are also expected due to the different manufacturing processes for AVs, their equipment and the appropriate infrastructure. Changes are also foreseen for the personnel working in passenger stations, freight storage facilities, terminals, maintenance and repair centres and other transport infrastructures. New job opportunities will be created in the fields of informatics, data management and telecommunications. It is possible that these new opportunities will increase the attractiveness of the transport sector for younger persons and women (European Commission, 2017, 2018b).
Personal data and security	The operation of AVs involves the collection of vast amounts of data, including data on travel choices and driving profiles, raising questions about personal data governance and usage. Researchers also highlight the potential vulnerability of AVs towards cyber-attacks and digital security (Parkinson et al., 2017).
Ethics and liability	The ethical dilemma about the deployment of autonomous mobility can be summarised into whether there should be *key algorithms that could result in life or death* decisions (Lin, 2016). During the course of an accident, a fully autonomous vehicle will have to decide on a course of action that may cause injury to its passengers or other road users. Based on current technological capabilities, this decision will be based on computer protocols, mirroring the existing legal and regulatory framework. It is however possible that this framework may not be able to sufficiently cover complex ethical problems with human-like sensitivity (Dogan et al., 2020; Goodall, 2014). In road accidents involving human-driven vehicles, the driver assumes primary liability as the person who controls the vehicle. In the case of AVs, being self-driven, the vehicle may take a part or the whole responsibility of an accident, as a form of product failure. This affects third parties which are responsible for the final product, such as vehicle and equipment designers, manufacturers and providers as well as software designers and providers, and operators. The development of the legal framework to proportionally assign liability and the related insurance costs to the involved parties remains a challenge (Taeihagh & Lim, 2018).
Digital accessibility and social exclusion	This concern arises out of the existing gaps between societal groups that are not able to navigate the digital world. Social exclusion due to limited skill and capability at accessing and using digital means, such as a smartphone, will remain due to transitional effects, economic effects and income, age and/or disability.

THE DATA ISSUE

Lack of Data During for Current Planning Purposes

The uncertainty from autonomous transport and its impacts on sustainable urban development derives from the lack of empirical data and previous experience. An increasing number of pioneer cities are implementing pilot tests and trials (as seen earlier in Table 7.1), and their experience, data and information are a valuable source for the estimation of potential benefits and drawbacks for other cities. This information has to be exchanged and analysed in the context of the specific characteristics and development goals of each city.

Studies are attempting, with the use of macroscopic models, to assess the potential impacts of autonomous mobility on different aspects of sustainable development under different penetration and implementation scenarios (Joint Research Center, 2018). There are also models which focus on specific cities, business cases for AV implementation or impacts, for example the study on the prospect of introducing autonomous demand responsive transit (DRT) in Cottbus, Germany (Bischoff et al., 2019), and the simulation of the impact of AVs on the relocation of households based on their effect on the reduction of the value of time, parking restrictions and congestion (Llorca et al., 2022). According to Berrada and Leurent (2017), the models which are usually used for the assessment of AVs vary according to the scope of the analysis. Various studies have also been conducted to estimate the impact of autonomous mobility on road safety (Litman, 2023). These models provide a useful insight to the planners and policymakers of a city. However, due to the scarcity of empirical data, they often rely on hypotheses about travel choices and behaviour, development trends and penetration rates that may not correspond to the local characteristics of all cities. Research is also directed towards the implementation of multicriteria analysis (Anastasiadou et al., 2021; Jiang et al., 2022), in-depth surveys (Cugurullo et al., 2020; Hewitt et al., 2019), workshops (Milakis et al., 2017) and other methods for assessing expert opinion or public acceptance regarding the introduction of autonomous transport systems in a city and the anticipated impacts.

The overview of international experience and empirical data in other domains may also be used as a basis for assumptions and foresight regarding the potential impact of autonomous mobility on cities and urban development. For example, the 'dominance' of the private car, especially after the second half of the 20th century, led to disruptive changes in mobility by increasing the flexibility, range and effectiveness of 'door-to-door' trips. According to Newman and Kenworthy (1996), the private car gradually became a main factor in shaping urban development, especially in the so-called Western world. The extensive use of private cars allowed for low-density housing, land use zoning and the decentralisation of activities, leading to the increase of the distance and frequency of trips. In the decentralised cities, public and active modes of transportation became less relevant, while car dependence eventually created congestion and negative impacts on sustainable mobility and sustainable urban development (Rode et al., 2017). In this context, Banister (2008) argues that the sustainable mobility paradigm should address the above challenges by strengthening the connection between transport and land use. The evolution of private car use and its impact on cities has been thoroughly examined and this experience can be adjusted to support assumptions on the possible impacts from the deployment of private autonomous cars in a city.

Moreover, automation in other transport modes, such as the automation of urban railways, introduced in London in 1967, can be used to support assumptions on the socio-economic effects of autonomous public transport. The impact of urban railways on spatial development and land use (Huang, 1996; Newman & Kenworthy, 1996) could be also used to make assumptions on the relevant impacts of autonomous Bus Rapid Transit (BRT) systems. Conclusions on the economic impact of automation in the transport sector can be drawn from the close examination of automation in other sectors. For example, the massive deployment of Automatic Teller Machines (ATMs) in the banking sector (Autor, 2015) can support basic assumptions about the impact of transport automation on jobs as well as on emerging business models. According to Bessen (2015), the deployment of ATMs resulted in the reduction of personnel for each bank branch but in the increase of the number of branches due to lower operational costs.

Big Data for Planning During the Wide-Scale Deployment of AVs

Connected AVs apply different sensors, cameras and wireless telecommunication technologies to collect real-time data about traffic conditions, road infrastructure, vehicle and engine metrics, trip and route characteristics, passenger characteristics, travel behaviour and other user information and environmental conditions (such as air quality, noise). They are able to collect and exchange real-time information with drivers and users of other modes, vehicles, infrastructure and any other entity interacting with the transport network (Vehicle to Everything communication). The progress in information and telecommunication technologies and the integration of the city's transport and telecommunication networks play a key role in this aspect (ERTRAC Working Group: 'Connectivity and Automated Driving', 2019).

In the future, the wide-scale deployment of AVs will produce vast amounts of data, which could be used for transport planning and traffic management purposes (Joint Research Center, 2017). They could be part of the city's big database, deriving from various sources, such as buildings, service providers, infrastructure, and social media, and used to support evidence-based urban planning (Lim et al., 2018). In order to take advantage of this opportunity, the policy and planning bodies should be able to effectively use these data, based on the appropriate data governance framework for the safe access and efficient management of big data (EuroCities, 2018; Kerber, 2019). The design and implementation of the data governance framework should take into account a series of main challenges (Costantini et al., 2020, 2021; Innamaa et al., 2018):

- The personal information from users which are collected by AVs should be used and shared according to the rules and specifications of sensitive personal data protection.

- The access to and format of data from businesses (such as autonomous vehicle manufacturers and operators, transportation and telecommunication service providers and ICT companies) should be treated according to the regulations about competitive or proprietary information.
- In order for AV data to be useable for planning purposes, the corresponding datasets should be well documented and tested, data collection methods should be thoroughly reported and extensive metadata should be available.
- Collecting and managing the vast amounts of real-time data from AVs is a complicated and costly process which requires effective and efficient cooperation between the data owners, providers and re-users.

RECOMMENDATIONS FOR PLANNERS

The sustainable urban mobility principles have been implemented for over two decades, aiming at the promotion of public, shared, and active transport, the change in travel behaviour and the management of transport demand through targeted measures and policies which are adjusted to the local specificities and urban development goals. Integrated urban transport planning in the era of autonomous mobility should capitalise on these principles.

Planners should assess and monitor the mobility-related features, i.e. the emerging mobility patterns and trip characteristics, the volume and type of generated traffic, the complementarity or competition of the new system with existing public and active transport modes and, especially in the case of collective (shared or public) autonomous transport, the shift to the new system from conventional private cars. Specific and tangible targets should be set to dynamically assess the performance of an autonomous transport system. Regarding the adjustment of the city's transport infrastructure to accommodate an autonomous transport system, it is essential to assess the geometric and functional characteristics of the road network, the capacity and location of parking and refuelling/charging and multimodal stations (e.g. Park and Ride facilities) and the technological and technical features of the city's traffic management system.

Cross-scientific cooperation is essential in planning for the deployment of autonomous transport systems in a city. More specifically, the collaboration between experts in different fields is needed for the integration of the Smart City's physical and digital infrastructure of the transport, telecommunication, and energy networks. It is also important for transport and spatial planners to collaborate in order to assess the spatial impacts of the deployment of autonomous mobility. The transport and mobility features should be examined in relation to the land-use features of the areas covered by the new system, i.e. origins and destinations of autonomous trips and passing-through areas, and issues of accessibility and mobility equity should be addressed. The estimation of external impacts from autonomous mobility on energy consumption, the quality of the environment and the GHG emissions also requires cross-scientific cooperation with environmental engineers and climate experts.

In the above framework, the following recommendations are crucial from the planner's perspective:

- Implementation of the appropriate time horizon for impact assessment in relation to two main phases, that is the transition and the wide-scale deployment of AVs.
- Adoption of participatory planning approaches to take into account the mobility and wider socio-economic impacts on persons with reduced mobility, commuters, professional drivers and other professionals, service providers, and companies.
- Knowledge of the different requirements of private, shared and public autonomous transport systems and of the different ways that each system is expected to affect mobility, urban development and environmental conditions.

CONCLUSIONS

This chapter has provided an overview of potential key impacts of AV in relation to sustainable urban development in order to offer strategic-level recommendations to transport and urban planners. The possible impacts from the deployment of autonomous transport on sustainable urban development are characterised by:

- Breadth covering effects on mobility and accessibility conditions, land uses and the spatial organisation of activities, energy consumption and environmental footprint and wider socio-economic conditions.
- Diversity, as they may be positive or negative depending on the type of autonomous transport system (private, shared or public transport), the specific features of the city and the urban development priorities.
- Uncertainty, due to the shortage of empirical data and real-life experience from wide-scale implementation.

Nevertheless, it is necessary for planners to examine the potential of autonomous mobility, to exploit its advantages for the promotion of sustainable transport and to address its possible impacts on urban development. This derives from the fact that AV deployment by 2030 is prioritised by policymakers in many parts of the world. At the same time, cities play a central role in addressing global societal and environmental challenges, including the provision of a safe, affordable, accessible and sustainable urban transport system by 2030 (UN SDG11). Moreover, they contribute to the global efforts to control climate change by 2050. Thus, cities are in need of an integrated transport planning approach to assess the possible impacts of AV deployment in their mid-term and long-term scenarios for the promotion of sustainable urban mobility. In this effort, planners may capitalise on the experience of pioneer cities, the cross-scientific cooperation, the combined use of different scientific methods and the participatory planning approaches.

Finally, the deployment of AVs is linked to their ability to collect vast amounts of real-time data, which can be used for planning purposes. Towards this purpose, the appropriate data governance structures should be developed to cope with the protection of sensitive data, the sharing of the costs from data collection and the know-how of planning and policy making authorities to effectively manage big data.

REFERENCES

Anastasiadou, K. (2021). Sustainable mobility driven prioritization of new vehicle technologies, based on a new decision-aiding methodology. *Sustainability, 13*, 4760.

Anastasiadou, K., Gavanas, N., Pitsiava-Latinopoulou, M., & Bekiaris, E. (2021). Infrastructure planning for autonomous electric vehicles, integrating safety and sustainability aspects: A multi-criteria analysis approach. *Energies, 14*, 5269.

Autor, D. H. (2015). Why there are still so many jobs? The history and future of workplace automation. *Journal of Economic Perspectives, 29*(3), 3–30.

Banister, D. (2008). The sustainable mobility paradigm. *Transport Policy, 15*, 73–80.

Berrada, J., & Leurent, F. (2017). Modeling transportation systems involving autonomous vehicles: A state of the art. *Transportation Research Procedia, 27*, 215–221.

Bessen, J. (2015). Toil and technology. *Financial Development, 52*, 16–19.

Bischoff, J., Fuhrer, K., & Maciejewski, M. (2019). Impact assessment of autonomous DRT systems. *Transportation Research Procedia, 41*, 440–446.

Bosch, P. M., Becker, F., Becker, H., & Axhausen, K. W. (2018). Cost-based analysis of autonomous mobility services. *Transport Policy, 64*, 76–91.

Canitez, F., Thomopoulos, N., & Cantafio, G. (2018). *AV trials overview, WISE-ACT Workshop 1*, March 2018, Bratislava, Slovakia.

Costantini, F., Di Ciommo, F., Sanguino, T., Carballo-Calero, P., Klar, W., Shallauer, D., Nguyen-Luong, D., Dogan, E., Arora, P., Backhaus, W., Franco, D., Rupprecht, S., Susilo, Y., Shiko, V., & Thomopoulos, N. (2021). *Institutional and regulatory challenges of autonomous and connected transport*, WG1 Thematic Report, COST Action CA16222 WISE-ACT.

Costantini, F., Thomopoulos, N., Steibel, F., Curl, A., Lugano, G., & Kováčiková, T. (2020). Autonomous vehicles in a GDPR era: An international comparison. In D. Milakis, N. Thomopoulos, & B. van Wee (Eds.), *Advances in transport policy and planning* (pp. 191–213). Elsevier.

Cugurullo, F., Acheampong, R. A., Gueriau, M., & Dusparic, I. (2020). The transition to autonomous cars, the redesign of cities and the future of urban. *Urban Geography, 42*(6), 833–859.

Dogan, E., Costantini, F., & Le Boennec, R. (2020). Ethical issues concerning automated vehicles and their implications for transport. In D. Milakis, N. Thomopoulos, & B. van Wee (Eds.), *Policy implications of autonomous vehicles* (pp. 215–233). Elsevier.

ERTRAC Working Group: "Connectivity and Automated Driving". (2019). *Connected automated driving roadmap*. ERTRAC.

EuroCities. (2018). *Integrating transport automation in the urban system*. https://eurocities.eu/wp-content/uploads/2020/07/EUROCITIES_Statement_on_Transport_Automation_in_Urban_Areas_final.pdf

European Commission. (2016). *A European strategy on Cooperative Intelligent Transport Systems, a milestone towards cooperative, connected and automated mobility, COM(2016) 766 final*. European Commission.

European Commission. (2017). *Mobility and transport. Women in Transport – EU Platform for change*. https://ec.europa.eu/transport/themes/social/women-transport-eu-platform-change_en

European Commission. (2018a). *Autonomous vehicles & traffic safety*. European Commission, Directorate General for Transport. https://road-safety.transport.ec.europa.eu/system/files/2021-07/ersosynthesis2018-autonomoussafety-summary.pdf

European Commission. (2018b). *Automation in transport: How does it affect the labour force?* https://ec.europa.eu/transport/sites/transport/files/2018-11-20-automation-in-transport-background.pdf

European Commission. (2020). *Sustainable and smart mobility strategy – Putting European transport on track for the future. COM(2020) 789 final.* European Commission.

European Commission. (2021). *The New EU urban mobility framework. COM(2021) 811 final.* European Commission.

European Commission. (2022). *Connected and automated transport (CAT)*. https://trimis.ec.europa.eu/roadmaps/connected-and-automated-transport-cat

Fagnant, D. J., & Kockelman, K. (2015). Preparing a nation for autonomous vehicles: Opportunities, barriers and policy. *Transport Research Part A: Policy Practice, 77*, 167–181.

Fraedrich, E., Heinrichs, D., Bahamonde-Birke, F. J., & Cyganski, R. (2019). Autonomous driving, the built environment and policy implications. *Transportation Research Part A: Policy and Practice, 122*, 162–172.

Gavanas, N. (2019). Autonomous road vehicles: Challenges for urban planning in European cities. *Urban Science, 3*(2), 61.

Goodall, N. J. (2014). Machine ethics and automated vehicles. In G. Meyer & S. Beiker (Eds.), *Road vehicle automation* (pp. 93–103). Springer International Publishing.

Greenblatt, J. B., & Shaheen, S. (2015). Automated vehicles, on-demand mobility, and environmental impacts. *Current Sustainable/Renewable Energy Reports, 2*, 74–81.

Gridsted, T. S., Christensen, T. H., Freudendal-Pedersen, M., Friis, F., & Hartmann-Petersen, K. (2022). The urban governance of autonomous vehicles – In love with AVs or critical sustainability risks to future mobility transitions. *Cities, 120*, 103504.

Hawkins, J., & Habib, K. N. (2019). Integrated models of land use and transportation for the autonomous vehicle revolution. *Transport Reviews, 39*, 66–83.

Heinrichs, D. (2016). Autonomous driving and urban land use. In M. Maurer, C. Gerdes, B. Lenz, & H. Winner (Eds.), *Autonomous driving* (1st ed.). Springer.

Hewitt, C., Amanatidis, T., Politis, I., & Sarkar, A. (2019). Assessing public perception of self-driving cars: The autonomous vehicle acceptance model. *ACM IUI 2019 Conference*, Los Angeles.

Huang, H. (1996). The land-use impacts of urban rail transit systems. *Journal of Planning Literature, 11*(1), 17–30.

Iacobucci, R., McLellan, B., & Tezuka, T. (2018). Modeling shared autonomous electric vehicles: Potential for transport and power grid integration. *Energy, 158*, 148–163.

iMOVE Australia. (2022). *Smart mobility projects and trials across the world.* https://imoveaustralia.com/smart-mobility-projects-trials-list/

Innamaa, S., Smith, S., Barnard, Y., Rainville, L., Rakoff, H., Horiguchi, R., & Gellerman, H. (2018). *Trilateral Impact Assessment Framework for Automation in Road Transportation. Version 2.0.* EU-US-Japan ITS Cooperation.

International Transport Forum. (2017). *Managing the transition to driverless road freight transport.* ITF/OECD.

Jiang, L., Chen, H., & Chen, Z. (2022). City readiness for connected and autonomous vehicles: A multi-stakeholder and multi-criteria analysis through analytic hierarchy process. *Transport Policy, 128*, 13–24.

Joint Research Center. (2017). *The R-Evolution of Driving: From connected vehicles to coordinated automated road transport (C-ART).*

Joint Research Center. (2018). *An analysis of possible socio-economic effects of a cooperative, connected and automated mobility (CCAM) in Europe.*

Kaplan, L., Rupprecht, S., Grosso, M., Thomopoulos, N., Backhaus, W., Alonso Raposo, M., Ciuffo, B., & Franco, D. (2023). Ensuring strong public support for automation in the planning process: From engagement to co-creation. In G. Meyer & S. Beyker (Eds.), *Road Vehicle Automation 9.* Springer International Publishing.

Kerber, W. (2019). Data governance in connected cars: The problem of access to in-vehicle data. *JIPITEC Journal of Intellectual Property, Information Technology and E-Commerce Law, 9*(1), 310–331.

Lim, C., Kim, K. J., & Maglio, P. P. (2018). Smart cities with big data: Reference models, challenges, and considerations. *Cities, 82*, 86–99.

Lin, P. (2016). Why ethics matters for autonomous cars. In M. Maurer, J. Gerdes, B. Lenz, & H. Winner (Eds.), *Autonomous driving. Technical, legal and social aspects.* Springer Open.

Litman, T. (2023). Traffic safety and public health. In T. Litman (Ed.), *Autonomous vehicle implementation predictions. Implications for transport planning* (pp. 13–14). Victoria Transport Policy Institute.

Llorca, C., Moreno, A., Ammar, G., & Moeckel, R. (2022). Impact of autonomous vehicles on household relocation: An agent-based simulation. *Cities*, *126*, 103692.

Lu, Q., Tettamanti, T., Hörcher, D., & Varga, I. (2020). The impact of autonomous vehicles on urban traffic network capacity: An experimental analysis by microscopic traffic simulation. *Transportation Letters. The International Journal of Transportation Research*, *12*(8), 540–549.

Martens, K., Beyazit, E., Henenson, E., Thomopoulos, N., Milakis, D., Mladenovic, M., Pudane, B., van Wee, B., Di Ciommo, F., Curl, A., Cugurullo, F., Dimitrova, E., & Negulescu, M. (2021). *WG2 thematic report: Autonomous and connected transport as part of an inclusive transport system*. COST Action CA16222 WISE-ACT.

Metz, D. (2018). Developing policy for urban autonomous vehicles: Impact on congestion. *Urban Science*, *2*(2), 33.

Meyer, J., Becker, H., Bosch, P., & Axhausen, K. W. (2017). Autonomous vehicles: The next jump in accessibilities? *Research in Transport Economics*, *62*, 80–91.

Milakis, D., Snelder, M., van Arem, B., van Wee, B., & de Almeida Correia, G. H. (2017). Development and transport implications of automated vehicles. *European Journal of Transport and Infrastructure Research*, *17*(1), 63–85.

Milakis, D., Thomopoulos, N., & van Wee, B. (2020). *Policy implications of autonomous vehicles*. Elsevier.

Newman, P. W., & Kenworthy, J. R. (1996). The land use-transport connection. An overview. *Land Use Policy*, *13*(1), 1–22.

Nikitas, A., Thomopoulos, N., & Milakis, D. (2021). The environmental and resource dimensions of automated transport: A nexus for enabling vehicle automation to support sustainable urban mobility. *Annual Review of Environment and Resources*, *46*, 167–192.

Nourinejad, M., Bahram, S., & Roorda, M. J. (2018). Designing parking facilities for autonomous vehicles. *Transportation Research Part B: Methodology*, *109*, 110–127.

Parkin, J., Clark, B., Clayton, W., Ricci, M., & Parkhurst, G. (2018). Autonomous vehicle interactions in the urban street environment: A research agenda. *Proceedings of the Institution of Civil Engineers. Municipal Engineer*, *171*(1), 15–25.

Parkinson, S., Ward, P., Wilson, K., & Miller, J. (2017). Cyber threats facing autonomous and connected vehicles: Future challenges. *IEEE Transactions on Intelligent Transportation*, *18*(11), 2898–2915.

Rode, P., Floater, G., Thomopoulos, N., Docherty, J., Schwinger, P., Mahendra, A., & Fang, W. (2017). Accessibility in cities: Transport and urban form. In G. Meyer & S. Shaheen (Eds.), *Disrupting mobility. Impacts of sharing economy and innovative transportation on cities* (pp. 239–273). Springer.

Rupprecht Consult. (2019a). *Road vehicle automation in sustainable urban mobility planning, practitioner briefing, European Platform for Sustainable Mobility Plans*. https://www.eltis.org/sites/default/files/road_vehicle_automation_in_sustainable_urban_mobility_planning_0.pdf

Rupprecht Consult. (2019b). *Guidelines for developing and implementing a sustainable urban mobility plan* (2nd ed.). https://www.eltis.org/mobility-plans/sump-guidelines

Ruso, A., van Ommeren, J., & Dimitropoulos, A. (2019). *The environmental and welfare implications of parking policies*. [Environment Working Paper No. 145. OECD].

Simpson, C., Ataii, E., Kemp, E., & Zhang, Y. (2019). *Mobility 2030: Transforming the mobility landscape*. KPMG.

Soteropoulos, A., Berger, M., & Ciari, F. (2019). Impacts of automated vehicles on travel behaviour and land use: An international review of modelling studies. *Transport Reviews*, *39*, 29–49.

Stead, D., & Vaddadi, B. (2019). Automated vehicles and how they may affect urban form: A review of. *Cities*, *92*, 125–133.

Taeihagh, A., & Lim, H. S. (2018). Governing autonomous vehicles: Emerging responses for safety, liability, privacy, cybersecurity, and industry risks. *Transport Reviews*, *39*(1), 103–128.

Thomopoulos, N., & Givoni, M. (2015). The autonomous car – A blessing or a curse for the future of low carbon mobility? An exploration of likely vs. desirable outcomes. *European Journal of Futures Research*, *3*, 14.

Transport & Environment. (2017). *Driverless cars increase congestion – but could cut massive parking times*. https://www.transportenvironment.org/news/driverless-cars-increase-congestion-%E2%80%93-could-cut-massive-parking-times

United Nations. (2015). *Paris Agreement*. United Nations. Climate Change. https://unfccc.int/sites/default/files/english_paris_agreement.pdf

United Nations. (2022). *Sustainable Development Goals (SDGs)*. Department of Economic and Social Affairs, Sustainable Development. Retrieved from 11. Make cities and human settlements inclusive, safe, resilient and sustainable: https://sdgs.un.org/goals/goal11

United Nations General Assembly. (2015). *Transforming our world: The 2030 Agenda for Sustainable Development*. United Nations. Department of Economic and Social Affairs. Sustainable Development. https://www.un.org/ga/search/view_doc.asp?symbol=A/RES/70/1&Lang=E

van Acker, V., Kyriakidis, M., Thomopoulos, N., Kolarova, V., Etzioni, S., Baburajan, V., … Abreu e Silva, J. (2021). *WG4 thematic report: Transport demand & user challenges of ACT*. COST Action CA16222.

van der Waals, J. F. (2000). The compact city and the environment: A review. *Tijdschrift voor Economische en Sociale Geografie*, *91*(2), 111–121.

Wadud, Z., MacKenzie, D., & Leiby, P. (2016). Help or hindrance? The travel, energy and carbon impacts of highly automated vehicles. *Transportation Research Part A: Policy and Practice*, *86*, 1–18.

Wegener, M., & Fürst, F. (1999). *Land use transport interaction: State of the art, Deliverable 2a of the project TRANSLAND (Integration of Transport and Land Use Planning)*. Institut für Raumplanung, Fakultät Raumplanung, Universität Dortmund.

Williams, B. (2021). *Automated Vehicles and MaaS: Removing the Barriers*. Wiley.

World Commission on Environment and Development. (1986). *Report of the World Commission on Environment and Development: Our Common Future*. United Nations. Department of Economic and Social Affairs. Sustainable Development. https://sustainabledevelopment.un.org/content/documents/5987our-common-future.pdf

World Economic Forum. (2018). *Reshaping urban mobility with autonomous vehicles. Lessons from the City of Boston*. http://www3.weforum.org/docs/WEF_Reshaping_Urban_Mobility_with_Autonomous_Vehicles_2018.pdf

CHAPTER 8

RESCUING TRANSPORT FROM INEQUITIES: HOW CAN AUTOMATED AND CONNECTED TRANSPORT CONTRIBUTE TO A MORE INCLUSIVE TRANSPORT SYSTEM?

Eda Beyazit[a], Emily Soh[b] and Karel Martens[b]

[a]University of the West of England (UWE), Bristol, United Kingdom
[b]Technion – Israel Institute of Technology, Haifa, Israel

ABSTRACT

The massive investment in, and development of, automated and connected transport (ACT) technology development has triggered much debate about this breakthrough technology's potential positive and negative impacts. Multiple studies have explored the potential direct implications for users in terms of road safety, 'productive' travel time, mobility of the elderly and physically less mobile persons, as well as indirect impacts such as reduced emissions and freed road space. Through a critical review of the literature on ACT deployment types and discussions with an expert working group on the wider impacts of ACT implementation, this chapter examines four distinct deployment types of ACT technology and their opportunities and threats in transitioning toward inclusive transport systems. Of the four types, we posit that ACT-based public transport has the greatest potential to contribute to a more inclusive mobility future. Examining the case of Singapore using policy documents, academic

Sustainable Automated and Connected Transport
Transport and Sustainability, Volume 19, 139–159

Published under exclusive licence by Emerald Publishing Limited
ISSN: 2044-9941/doi:10.1108/S2044-994120240000019008

literature and interviews with representatives of public and private sectors and academia, the chapter draws policy recommendations for governance toward more inclusive ACT innovation and deployment.

Keywords: Inclusive mobility; transport inequities; sustainability; automated transport; connected transport; ACT-based deployment types; public transport

INTRODUCTION

The role of ACT has been widely discussed in the last decade as a potential remedy to the problems cities and transport systems face, such as traffic congestion, air pollution, accidents and car-oriented development. This chapter explores the possible contribution of ACT in making the current surface transport system more inclusive and, thus, more sustainable in its broadest sense. For this, we take a people-centered perspective. This perspective asks whether people – in all their diversity – receive adequate service from all transport modes jointly to meet fundamental human needs and wants (Martens, 2017). Hence, the core question is: whose service is being improved by the introduction of ACT in its various forms?

Drawing on the extensive literature on transport equity and transport-related social exclusion (Lucas, 2012; Martens & Lucas, 2018), we will focus our exploration on the following poorly served population groups: low-income households, people with (motor, sensory or cognitive) impairments, women, children and youth. In addition, we take geography into account, as residential location also strongly shapes a person's mobility and accessibility. The literature shows that a substantial share of these groups is currently poorly served because they have no or limited access to a (reliable) car for financial, legal or ability-related reasons (Mattioli et al., 2017), while the transport means and systems available to them often do not provide a high level of mobility and accessibility. The lack of access to a car is not inherently problematic, but decades of car-oriented investments and urban developments have made it *de facto* challenging to access destinations without a car in much of the developed world. ACT has the potential to enhance inclusivity in transportation (Blyth et al., 2016; Horizon 2020 Commission Expert Group (2020); Herzogenrath-Amelung et al., 2015). However, a positive impact requires a deliberate focus on shaping an inclusive transport system, going beyond discussions on the proper introduction of ACT. Hence, in this chapter, we explicitly adopt the perspective of disadvantaged groups to scrutinize whether different forms of ACT-based transport enhance their mobility and accessibility.

The chapter is structured as follows. In the next section, we discuss how all transport modes are, to some extent, exclusionary by nature. This argument is also accurate for ACT. While ACT has the potential to enhance the suitability of regular (private) motorized vehicles for a broader user base, it cannot accommodate everyone in all situations. Hence, in the following sections, we discuss various possible ACT deployment types, differing in automation level and level of sharing.[1] Here, we use the term partial automation (in other words, low-level ACT) for Levels 2, 3 and 4, and full automation (high-level ACT) for Level 5 only (SAE, 2021). We limit the discussion on shared ACT to sequential sharing, as in 'not travelling with others', while collective mobility, i.e. sharing rides with unknown others, refers to ACT use in all forms of public transport (i.e., both regular and on-demand services). The focus lies in exploring how various deployment types may impact individuals with diverse socio-demographic characteristics, residing in different spatial contexts, and with different physical and cognitive abilities. This exploration concludes that ACT-based public transport, in various forms, has the most potential to contribute to an inclusive transport system. In the last sections, we explore how Singapore, one of the countries at the forefront of the ACT revolution, plans to enhance public transport services through ACT technologies. The chapter concludes with a broad reflection on the role of governments in implementing ACT, considering the potential futures that may or may not materialize if the introduction of ACT is not carefully managed.

The following sections are based on a critical review of the literature on different deployment types of ACT-based transport to extract their possible implications for different underserved population groups. We also engaged in multiple open-ended discussions with diverse experts within the WISE-ACT COST Action working group on Wider Impacts and Scenario Evaluation of Autonomous and Connected Transport. For the Singapore experience in introducing ACT-based technologies in public transport, we reviewed academic and grey literature, including policy documents on planning and transport and the websites of Singapore's government ministries, including the Land Transport Authority. Fifteen interviews were carried out with ACT-based transport practitioners from the public sector, private sector and academia in Singapore. From the literature and interviews, we derived the rationales, objectives, policies and actions undertaken by the government, often in collaboration with actors in academia and the private sector, in introducing ACT-based technologies in public transport.

THE EXCLUSIONARY CHARACTER OF TRANSPORT MODES

Transport planning and policy would have been much less complicated if one mode of transport fitted all. However, people vary greatly in their needs, wants, concerns, constraints and preferences. All transport modes exclude some people. For instance, walking or cycling a considerable distance may not be an issue for most people, while for others, even a short walk may be troublesome (Curl & Fitt, 2018; Park & Chowdhury, 2018). Likewise, using public transport may be challenging for people with physical and mental impairments despite intervention

---

[1]This chapter, including the four different deployment types, are partially an outcome of the discussions throughout the WISE-ACT COST Action working groups and to some extent, the WISE-ACT survey that proposed different scenarios including private and shared ACT vehicles. Since some of the ACT deployment types we distinguish are likely to be introduced in parallel, we use the term deployment types instead of scenarios.

efforts toward inclusive designs (Martens, 2018). This exclusionary character of transport modes often leads people to choose more comfortable, practical (door-to-door) and personal modes of traveling, such as private cars, whose adverse impacts on sustainability have been discussed in transport and urban research for decades (Banister, 1999; Newman & Kenworthy, 1999). Yet, private cars are only available to some due to financial barriers and social norms, even in more affluent countries (Uteng, 2009). Even if financial barriers were taken away, especially for young people and individuals with various impairments, driving a conventional car would not be possible. This is also the case for some share of older adults, who may face the prospect of having to stop driving as their abilities decline as they age (Dellinger et al., 2001; Rosenbloom, 2010).

While the current debates lead to an understanding that ACT holds the potential of eliminating the challenges of active mobility for all as well as some of the exclusionary characteristics of the car and of public transport services, it is unlikely that all people will be able to use a fully automated ACT vehicle. For instance, approval would be needed from caretakers for children and youth to use ACT for reasons of safety, personal security, and privacy (Koppel et al., 2021). Likewise, ACT may exclude some adults with low technological literacy and/or without access to a bank account or credit card, as using ACT is likely to require digital skills (Milakis & van Wee, 2020). ACT technology may widen the already existing digital divide, potentially exacerbating socio-spatial inequalities (Thomopoulos & Givoni, 2015). Other conditions, such as distrust of technology that is usually driven by the fear that Artificial Intelligence (AI) will be unable to prevent accidents, that personal data will be leaked due to cyber-attacks, or the AI will fail to perform its main tasks (Cugurullo & Acheampong, 2023), may also impede the ACT use among some share of the population.

Nevertheless, the primary obstacle to the widespread adoption of private ACT will predominantly be financial. Even in affluent nations, a significant portion of adults currently lack the financial means to afford car ownership (e.g., Allen & Farber, 2021) or even a subscription to a car-sharing system. Additionally, some share of households grapple with the financial burden of forced car ownership and often face infrequent or unreliable access to a vehicle (Currie & Senbergs, 2007; Mattioli, 2017). Even if ACT vehicles can become more affordable in the long term, global car ownership trends suggest that ownership of an ACT vehicle will not be within easy reach for a substantial share of the population. Likewise, shared ACT services might not appeal to individuals with lower incomes, given the experiences with existing car-sharing services which are primarily utilized by more affluent citizens (Kim, 2015; Kumar Mitra, 2021; Polydoropoulou et al., 2021). Therefore, we can argue that the wider adoption of ACT technology will be affected significantly by financial barriers, as discussed in the following sections under different ACT deployment types.

This exclusionary character of ACT (and other transport modes) should be a major concern for governments as they decide on their role and responsibilities in shaping ACT future (Shiftan et al., 2021). Hence, this fundamental understanding shapes the analyses that follow.[2]

---

[2]For a more extensive discussion on possible deployment types please see Martens et al. (2021).

DEPLOYMENT TYPE 1: PRIVATELY-OWNED PARTIALLY AUTOMATED ACT VEHICLES

While ACT technology is advancing swiftly, it is expected that the technology will still necessitate a driver to assume control of the vehicle at certain times in the foreseeable future. As long as this requirement exists, the use of private ACT vehicles will be possible only for individuals who have a valid driving license. Hence, partially automated ACT vehicles are most likely to enter the market as regular, mostly privately-owned, vehicles (even if these vehicles might also be used in public transport and for sharing services). Against this backdrop, we investigate the potential contribution of privately-owned, partially-automated ACT vehicles to fostering an inclusive transport system.

The introduction of limited yet reliable driver support may have some positive benefits. These driver support technologies are likely to enhance the mobility and accessibility of some drivers who find driving for long hours, over longer distances or in particular traffic situations stressful. ACT-technologies may support the elderly to continue driving for a longer period of time, as the technological support offered by higher automation levels (Level 3 or 4) may enhance their driving confidence (Parker et al., 2001). They may also benefit women drivers who tend to self-regulate more than their male counterparts (Gwyther & Holland, 2012).

On the other hand, non-users may be negatively affected by the deployment of private ACT vehicles. Downsides of ACT technologies might encompass heightened car usage, more precarious driving habits (as a result of behavioral adjustments), and added risks during take-over requests, potentially leading to increased risks for vulnerable road users, such as pedestrians, cyclists, and users of other small vehicles (e.g., mobility scooters). However, ACT technologies also present opportunities. Safety advantages could be particularly realized if external control over vehicle speed were implemented, akin to current regulations regarding shared e-scooters and e-bikes in various cities across the world (Tice, 2019). Maintaining speed limits is a proven method for decreasing crashes and creating safer neighborhoods, with benefits for especially young children, youth, older adults and individuals with impairments.

From a sustainable urban development perspective, ACT technology may lead to increased car use, which may trigger a new wave of car-oriented development and possibly urban sprawl. Following the domino effect, new developments, especially in suburban and peri-urban areas, may grow without proper public transport, possibly leading to a reduction in accessibility to opportunities for people without access to a car. Despite the efforts to reduce car ownership in cities worldwide, ACT technologies may increase the attractiveness of owning a car and could thus reverse the achievements of sustainable mobility thinking in recent decades.

This discussion brings us to the role of governments in harnessing the potential of ACT technology for a more inclusive transport future, as the introduction of low-level ACT vehicles may have detrimental equity impacts if left to the market. Governments could mitigate some equity implications by introducing some level of protection to unprotected road users through external speed control. Governments may make this obligatory and can introduce speed control on every new vehicle with ACT technology (Levinson, 2018). This would, in the long term, calm the traffic and support healthier, quieter, safer neighborhoods and cities. In addition, governments could control urban sprawl and require adequate public transport services for new urban developments, although current planning practices show that public authorities have been only modestly successful in this respect.

In conclusion, the introduction of privately-owned partially automated ACT-vehicles is unlikely to represent much progress toward an inclusive transport system. Some small share of users may experience enhanced mobility and, thus accessibility for certain trips. Unprotected street users, i.e. pedestrians, cyclists, and users of other light-weight vehicles, may benefit if ACT-technology is leveraged to control vehicle behavior in mixed traffic. The latter, however, will not occur automatically with the arrival of low-level ACT and would require government intervention.

DEPLOYMENT TYPE 2: PRIVATELY-OWNED FULLY AUTOMATED ACT VEHICLES

Fully automated vehicles that do not require a driver, whatever the circumstances (i.e., Level 5 automation), remain the long-term promise of ACT technology. While fully automated vehicles are tested in different cities across the globe, their integration into regular mixed traffic will take a long time. Yet, full automation may take over the streets at some point in different built environment settings. Here, we explore whether fully automated privately-owned cars can contribute to a more inclusive transport system.

The introduction of fully automated private ACT vehicles is expected to expand the demographic that can enjoy the advantages of car-based mobility and accessibility. In some way, this can be considered an advance toward a more inclusive transport system, particularly in light of the car-centric land use patterns in many regions across the world. However, many population segments are less likely to enjoy enhanced mobility and accessibility, most notably youth, individuals with mobility challenges, older adults, households with lower incomes, and women.

As discussed earlier, low-income groups already have limited resources to purchase and maintain private cars, let alone own fully automated vehicles with most certainly higher market values. The possibility of imposing full automation on all private vehicles for safety reasons (Sparrow & Howard, 2017), should the technology be widespread in the future, may substantially increase the costs related to car ownership and may thus potentially increase the gap in private car ownership between low and higher-income groups (Mobile Lives Forum, 2021). Even if it turns out that ACT vehicles are as affordable as traditional private cars, the presence of a multitude of advanced in-car technologies could raise the maintenance expenses for used cars, a factor particularly pertinent to low-income households (Currie & Senbergs, 2007). Therefore, insisting on a complete transition to fully automated private ACT vehicles may even increase the financial barrier for low-income households to own a (reliable) car, *ceteris paribus*.

For the younger populations who are financially dependent, the cost of ACT will likely be a challenge. In some households, this may be mitigated by sequential car sharing among household members, with youth using ACT vehicles under "parental controls" (similar to internet and TV use). Such a model might enable the adoption of fully automated vehicles among some share of youth. However, shuttling vehicles back and forth between different users and destinations will be costly, especially in where distances between locations are substantial, such as in suburban, peri-urban and rural areas (Litman, 2020). Hence, sequential sharing may be less prevalent in such areas than is suggested in some studies. In addition, there may be privacy and safety concerns (Costantini et al., 2020; Kyriakidis et al., 2020) for minors using fully automated vehicles.

While sequential sharing within households could work well in high-density urban areas due to shorter distances between destinations, ACT may increase traffic and replace much of the short-distance walking and cycling trips in urban areas, bringing about unsustainable mobility consequences. To some extent, this can be addressed by policies that restrict car use, such as parking management schemes, congestion charging schemes and the like.

Fully-automated ACT vehicles could bring substantial benefits to individuals with motor, sensory or cognitive impairments, and to some share of the elderly who may gradually lose the ability to drive a regular car. In fact, it is estimated that fully automated vehicles may increase the demand among these populations, with estimates suggesting that total vehicle kilometers traveled could rise by as much as 14% (Harper et al., 2016). The benefits of ACT use will obviously only accrue for people who are technologically literate and can afford an ACT vehicle. Yet, it should also be noted that people with impairments often have relatively low incomes, suggesting that ACT vehicles will be financially out of reach for them unless ACT use would in some way be subsidized (Kavanagh et al., 2015).

Women constitute a fourth group warranting explicit discussion. Women currently have less access to cars than men, because of lower income levels, gendered patterns of car access, and a range of cultural norms (Beyazit & Sungur, 2019). Moreover, concerns over social safety restrict women's travel by other transport modes, such as public transport and cycling, especially in the evening and night hours and in particular areas. Although it is difficult to predict its impacts on social norms, fully automated ACT may increase the social safety of traveling for women, which is often perceived as a barrier to employment by women as well as their spouses, fathers, or sons in patriarchal societies. Furthermore, especially in middle- to high-income households, fully automated vehicles may free women from some chauffeuring tasks, thereby reducing the burden related to the "mobility of care" (De Madariaga, 2013).

In sum, the introduction of privately-owned fully automated ACT vehicles via the market may imply some progress toward a more inclusive transport system, but only for particular segments of the population. The benefits of improved mobility and accessibility are most likely to be reaped by a (substantial) share of people with a range of impairments (including some of the elderly), provided they have the financial means for private ACT vehicles and are sufficiently tech-savvy. Very modest benefits may be reaped by a small segment of youth. Given the expected costs of ACT-vehicles, low-income households have little to gain from a market-driven introduction of fully automated vehicles.

DEPLOYMENT TYPE 3: SEQUENTIAL SHARING OF FULLY AUTOMATED ACT VEHICLES

The third deployment type we explore is the sharing of fully automated vehicles. Fully automated ACT sharing may be realized on a large scale relatively faster than fully automated private ACT because its introduction is likely to be driven by large corporations. As is the case with today's car-sharing schemes, these companies will offer automated vehicles with a support service, which may take away some of the concerns regarding ACT use. Moreover, thanks to the recent proliferation of sharing systems, not only carsharing but also bicycling and e-scooters, and their adaptability across many countries, the sharing culture that is already in place among some share of the population may assist in the early adoption of ACT sharing. In this section, we thus explore how the sequential sharing of fully automated ACT vehicles can contribute to an inclusive transport system. Our focus is limited to sequential sharing, where users travel individually or in a group in ACT vehicle, but do not share their journey with people they do not know.

Compared to the first two deployment types, ACT-based sequential sharing would bring more actors into the ACT ecosystem (Shaheen et al., 2020). One important factor of these entities includes large companies with the financial capacity to provide a varied vehicle fleet that aligns with the changing needs and preferences of the population in a given service area. Providing a range of vehicles and car-sharing subscription schemes could make ACT use economically viable and attractive. Yet, three barriers might still limit the use of ACT-based sharing services and, thus, the mobility and accessibility benefits it may bring to a range of population groups.

The first barrier is cost. Research shows that using regular, non-ACT car-sharing services is very modest among low-income households, mainly because of the high out-of-pocket costs for each trip (Clark & Curl, 2016; Forth Mobility, 2020; Shaheen et al., 2017). While ACT-sharing services maybe cheaper, given possible efficiency gains in fleet maintenance thanks to automation, out-of-pocket costs for each trip are likely to remain substantial, which will limit its use among low-income households. These out-of-pocket costs are also likely to be an inhibiting factor for youth, as discussed above, suggesting that uptake of ACT-based sharing among youth is likely to be limited to children from higher income groups.

ACT-based sharing services may be attractive for a substantial share of people with sensory impairments, such as people with visual impairments. This group may indeed substantially benefit from ACT-based sharing services, provided the entire trip is taken care of, from the point of ordering an ACT-vehicle, to finding the ordered vehicle in the street, to wayfinding from the moment of exiting the vehicle. For people with other types of impairments, shared ACT-services may still present barriers related to the design of the vehicle and the streetscape. Some individuals may require assistance with getting in and out of a fully automated vehicle, especially if they are traveling with a wheelchair or other mobility devices. Companies that provide sequential sharing may support these specific users with dedicated personnel, either within vehicles or remotely, through differentiated vehicle designs based on needs (Kyriakidis et al., 2020). This underscores that the introduction of shared ACT should be accompanied by regulatory regimes regarding vehicle design and support services to guarantee inclusive ACT deployment. The need for such regulations is underscored by experiences with ride-hailing services (Young & Farber, 2021).

The third obstacle concerns the spatial and temporal availability of shared ACT vehicles. As sharing thrives in high-demand areas, instant access to ACT-based carsharing services is less probable in locations and during periods of low demand, such as low-density suburbs and small towns, and potentially in low-income neighborhoods. While the possibly lower costs of operating and maintaining a shared ACT-fleet in comparison to a regular car fleet may (slightly) increase spatial and temporal availability, this will certainly not guarantee universal availability. Indeed, while in high-demand areas an abundance of shared ACT vehicles may typically be expected, guaranteeing availability within minutes, the longer distances, in combination with a lower density of shared ACT vehicles, will result in a longer lead time in low-demand areas. This may lead to the well-known vicious cycle in the suburbs, peri-urban and rural areas, with people preferring private (ACT-) vehicle ownership, leading to less demand for shared ACT vehicles, potentially leading to less supply and longer waiting times, which in turn would further encourage private car ownership.

This brief analysis suggests that shared ACT-services, while theoretically holding some promise for the environment, sustainability, and livable cities (Marsden et al., 2019; Nikitas et al., 2021), have, at best, a modest contribution to an inclusive transport system. This points to a need to include the criteria of social sustainability – which includes addressing accessibility and existing social needs (López et al., 2019) – in the governance of ACT deployment. From the groups poorly served by the current transport system, only some segments of people with impairments and some lower-income people may benefit from a shared ACT future, and only if governments guarantee and enforce the universal design of ACT-vehicles and related digital apps. Other groups, especially the majority of low-income households and people living in low-density areas, are unlikely to reap many direct benefits from ACT-based sharing. A shared future could, however, have an indirect positive contribution toward a more inclusive transport system. Potentially, ACT-sharing could lead to an increase in car-free and car-lite lifestyles in higher-density areas, with people combining ACT-sharing with walking, cycling and public transport (see the section on Singapore's transport strategy using walk-cycle-ride journeys). Such a development could strengthen an inclusive urban design and increase investments in 'active' and public transport modes, which in turn would improve mobility and accessibility for people with no or limited access to a (private, ACT) car, with possible impacts also beyond highly urbanized areas.

DEPLOYMENT TYPE 4: ACT-BASED PUBLIC TRANSPORT

In this section, we explore how ACT-based public transport, that is ACT services open to simultaneous use by non-related people, can contribute to an inclusive transport system. Currently, low levels of automation (Level-2 and Level-3) are already widely available and offer features like adaptive cruise control, lane keeping, and acceleration and braking assistance. These features can reduce fuel consumption and vehicle wear (Wadud et al., 2016) while enhancing comfort for passengers (Guo et al., 2020, 2021). Level-4 automation may offer additional cost savings. Yet, it is fully automated public transport (Level-5) that offers entirely new possibilities, warranting further discussion here.

The primary benefit of fully automated public transport, unlike partial automation, is a potential reduction in labor costs. These presently make up 40–70% of public transport operational costs in developed countries (Tirachini, 2020). Eliminating the need for a driver or operator goes beyond cost savings, providing the opportunity for a radical reimagining of public transport. The introduction of full automation offers the opportunity to develop an entirely new mix of regular schedule-based public transport and on-demand services (Martin, 2019). Even within existing public transport budgets substantial service expansions can be achieved through automation, notably in terms of increased frequencies, expanded service hours, and increased geographical coverage. These regular services may be complemented with a range of on-demand services, especially in low-density neighborhoods, peripheral areas or areas where the topography poses challenges for even short walks, especially for the elderly and individuals with impairments (Patel et al., 2021). In contrast to current offerings, full automation will make it possible to offer such on-demand services during the same (expanded) service hours as regular public transport. This implies that these on-demand services can serve as reliable feeders to regular public transport "trunk" lines or stations, going beyond on-demand feeder services currently offered in some locations (Bösch et al., 2018). Such feeder services are also part of Singapore's policy, where on-demand services complement mass public transit (see below). Beyond enhancing frequencies, service hours and coverage, full automation can also enhance reliability due to automated control of on-schedule operation, thus reducing transfer times and substantially enhancing passenger comfort. All these advancements will improve the mobility and accessibility of a large share of the population that is poorly served by the existing transport system (Etminani-Ghasrodashti et al., 2021). The possible high quality of a fully automated public transport system will not only increase ridership among these population segments but is also likely to attract others to start riding public transport, potentially setting off a positive cycle in which service improvements and increased ridership strengthen each other.

Clearly, a fully automated public transport system may also have disadvantages, most notably in terms of job losses in the short-term. Yet, new jobs would be expected to emerge with new demand in AV-related services and manufacturing. Indeed, the unemployment rate gap in the early years of ACT deployment may be lower than anticipated, as suggested by Groshen et al. (2019).

Based on the discussion of the four deployment types, Table 8.1 provides an overview of possible government interventions that may steer the different deployment types toward an inclusive transport system (see also Cohen & Cavoli, 2019). It can be noted that for privately-owned ACT vehicles, there are substantially fewer policy tools available that can be applied to achieve broad-based inclusivity, whereas, in the case of ACT-based public transport, this possibility is maximized.

SINGAPORE CASE: ACT-BASED PUBLIC TRANSPORT IN ACTION

The overview of deployment types suggests, perhaps unsurprisingly, that ACT-based public transport may contribute to a more inclusive transport system. If this thesis is accepted, the question is then: how can governments encourage the adoption of ACT-technologies in existing and new public transport systems? Hence, in this section, we explore how Singapore, one of the leading countries in ACT-readiness and ACT-experimentation, is seeking to leverage ACT-technology to enhance its public transport system.

Singapore's Transport Vision

A 2018 report by McKinsey finds that Singapore's public transport is among the top-ranked cities, scoring highly in all five key dimensions of availability, affordability, efficiency, convenience and sustainability (Knupfer et al., 2018). Singapore's pursuit of a high standard of public transport complements its long-standing policy of restricting private car ownership and use. Coordinated planning of land use and public transport infrastructure has served to optimize density around such infrastructure. This helped to secure high usage of public transport services and has resulted in compact and efficient land use. Singapore's public transport provision consists of a well-connected and growing metro network – the Mass Rapid Transit (MRT) – and a comprehensive bus network. In addition, taxi, ride-hailing and ride-sharing services are provided by the private sector and regulated by the government.

Singapore's vision for the future transport system is laid out in the Land Transport Master Plan 2040 (LTA, 2019). This plan envisions a convenient, well-connected and fast transport network through inclusive infrastructure. This vision is expressed through the goals of "20-minute towns" and a "45-minute city," jointly referring to a transport and land use system that allows people to complete most journeys in those timeframes using walk-cycle-ride combinations. These two goals are part of the plan's "transport for all" vision, which focuses on inclusive accessibility to everyone (LTA, 2019). It is upon this foundation and emphasis on public transport that Singapore considers its ACT transport strategies.

ACT-technologies are being considered as part of this vision for a car-lite future and as one of the ways to address demographic, urban and transport challenges facing Singapore. These challenges include an aging population, a shortage of operators for buses and other vehicles, and the need for more efficient use of land

Table 8.1. Possible Policy Interventions to Steer ACT Toward Inclusivity (*Source*: Compiled by Authors).

Areas of Intervention	Policy	Privately Owned Partially Automated ACT Vehicles	Privately Owned Fully Automated ACT Vehicles	Sequential Sharing of Fully Automated ACT Vehicles	ACT-Based Public Transport
Vehicle and system design accessibility	Speed limit for ACT vehicles	√	√	√	√
	Universal design of vehicles			√	√
	Inclusion of additional paratransit fleets			√	√
	In-vehicle security measures (for minors, emergency situations)			√	√
	User design interfaces on apps and vehicles (including for blind, hearing-impaired users, etc.)	√	√	√	√
	Ability to order vehicles without using digital technology (e.g. through telephone)	√	√	√	√
Urban design accessibility	Universal Design for infrastructure (drop-off points, stations, curb design, unobstructed pavements, etc.)	√	√	√	√
Spatial planning accessibility	Restricted use in dense urban areas	√	√	√	
	Ease of transfers: Incentivize connectivity to (and across) public transport modes	√	√	√	√
	Ensure provision of service to urban peripheries, low-income neighborhoods, etc.			√	√
Economic tools for accessibility	Utilizing cost savings to improve service quality, especially for inclusivity				√
	Enhance affordability through (targeted) subsidies			√	√

reserved for transport, given land scarcity. The aging population requires a rethink about how public transport can be made accessible for all, both in terms of physical distance to public transport stations and stops, as well as in vehicle design.

Given these challenges, high-quality public transport is to be enhanced by automated mobility technology to achieve greater efficiency and connectivity, thereby delivering greater accessibility and convenience for everyone. ACT is expected to improve the already high-quality public transport system in several ways: optimization of schedules for service quality improvement; broadening choices for travelers based on their needs and preferences for travel modes, cost, and environmental footprint of their trips; enhancing accessibility through greater flexibility of timings and drop-off and pick-up locations, shorter waiting and travel times, extending reach and service timing of transport services; and addressing supply-demand mismatches of fixed schedule services on low demand routes such that service is available only when needed. Given these potentials, Singapore is investing substantially in ACT-based public transport (see LTA website: http://www.lta.gov.sg/).

Promoting ACT-Based Public Transport

Singapore sees the potential of ACT technologies in a range of sectors, including last-mile freight delivery and surveillance. Among the many initiatives is also a concerted effort to explore and test the potential of ACT-based public transport. Singapore has developed an AV deployment roadmap with three phases for the introduction of automated (public) transport: test beds, town-deployment, and country-wide deployment. The roadmap is supported by abroad ACT-ecosystem that seeks to gradually introduce ACT-based public transport. As a key part of the ACT-ecosystem, Singapore's Land Transport Authority (LTA) has active cross-sector experimentations supporting the development of ACT-based public transport. These include:

- Collaborations with industry partners to test shared, on-demand, first-and-last-mile and intra-town self-driving transport concepts.
- Partnership with the Energy Research Institute at the Nanyang Technological University (ERI@N) to develop automated bus technologies, including self-driving bus trials for fixed and scheduled services (LTA, October 19, 2016).
- Collaboration with ST Kinetics to develop automated vehicle technologies for two 40-seater electric buses to serve fixed and scheduled services for intra- and inter-town travel (LTA, April 10, 2017).

Guidelines and Regulations

A sound legal framework is considered essential in creating and maintaining trust among the wider public regarding ACT-trials. Hence, the trialling of automated vehicles in Singapore, whether in the public transport system or otherwise, needs to adhere to mandatory regulations such as having comprehensive insurance coverage; having a "black box" data recorder that collects video footage and other data that can assist in investigations in the event of accidents; and real-time transmission of location and operational status to LTA's ACT vehicles monitoring system (LTA website).

Singapore has also established a comprehensive set of national guidelines called the Technical Reference (TR 68) to support the safe deployment of automated vehicles. First published in 2019 and revised in 2021, the guidelines cover four aspects: basic behavior, safety, cybersecurity principles and assessment, and vehicular data types and formats (LTA, September 3, 2021).

Trials

Singapore has conducted various trials with ACT-based public transport, using a range of vehicles in different settings such as business parks, university campuses, and leisure destinations. Notable examples of trials with regular passengers include:

- Fully automated shuttles with a capacity for 12 people were trialled in 2019 for a year at the National University of Singapore. The shuttles have built-in ramps and can accommodate a wheelchair.
- On-demand driverless minibuses and shuttles were trialed in 2019 for three months on the island of Sentosa.
- Trials for on-demand 10- and 26-seater driverless buses for fare-paying travelers were carried out in 2020 in the Singapore Science Park 2 and Jurong Island.

Larger scale trials for dynamically routed and on-demand busses and shuttles are planned for three areas: Punggol, Tengah and the Jurong Innovation District. These are, respectively, an established residential district, a greenfield residential district, and a business park district. Since 2019, the testing area has been gradually expanded to cover the entire western region of Singapore, consisting of more than 1,000 km of public roads.

Research

A final component of Singapore's ACT-ecosystem consists of (academic) research. Singapore's focus on applying automated mobility technology in public transport has generated a stream of research. Some of these studies have analyzed the interplay and combination of ACT-based public transport with other modes toward better coordination across transport system services, including first and last-mile connectivity (Chong et al., 2011; Mo et al., 2021). For instance, Shen et al. (2018) find that preserving existing high-demand bus routes while replacing low-demand bus routes with shared AVs potentially enhances service quality and efficiency, consumes fewer road resources, and is financially sustainable. Another subset of studies focuses on rider-oriented design aspects of ACT-based public transport (Cornet et al., 2019). Ongel et al. (2018) examine the use of an automated semi-rapid transit system with platooning capabilities that also incorporates Universal Design Standards to improve the service quality and efficiency of the transit systems. The study by Lim et al. (2021) explored the benefits of a "virtual companion" communication system for individuals with autism and disabilities and proposed ways on how it can be used in ACT-based public transport. Yet another subset of studies on public preference and acceptance of ACT-based public transport (Chng et al., 2021; Kurniawan et al., 2021) provides directions for designing ACT systems that are more responsive to user requirements.

Lessons Learnt

The Singapore efforts underscore that more development and testing are necessary before ACT-based public transport can be offered on a wider scale. Singapore holds an interesting balance between maintaining flexibility toward ACT-based transport innovation and experimentation and being intentional in directing some of these efforts to respond to its transport, demographic and societal challenges. Its public goals of inclusivity and accessibility through a high standard of public transport provision and a "car-lite" future shape its adoption of ACT-transport. Arguably, this focus on public transport, in effect, prioritizes a range of public values such as equity accessibility and civic life (all relating to social sustainability), and also environmental sustainability, well-being and quality of life (Docherty et al., 2022; Soh & Martens, 2023). This Singapore study provides an example of placing public goals front and center in considering future mobilities rather than enabling the introduction of market-driven technologies that mostly benefit people with the ability to pay.

DISCUSSION AND CONCLUSIONS

In this chapter, we examined whether ACT technologies could improve mobility and accessibility for population segments typically underserved by the current transport system. Utilizing a human-centric approach, we explored if and how these segments might benefit from different ACT technologies. We distinguished between four ACT deployment types: privately-owned partially automated ACT-vehicles, privately-owned fully automated ACT-vehicles, sequentially shared ACT-vehicles, and ACT-based public transport. Finally, we have reviewed the policies undertaken in Singapore as good practices that help the city attain its goal of a more inclusive transport system.

We argue that the ACT deployment type is definitive in generating sustainability and inclusivity benefits. Partial and full automation of private vehicles may generate more environmental burdens by encouraging car use and leading to more car-centric built environments. At the same time, especially full automation may enhance inclusion by making private car use more available to individuals who lack the ability to drive a conventional car. However, the expected costs of privately-owned ACT vehicles and the potential legal requirements regarding their use will constrain the benefits to a limited segment of individuals poorly served by the current transport system. Hence, both partially and fully automated private ACT vehicles are unlikely to serve middle to low-income populations living in less-developed parts of the world.

While fully automated sequential sharing services could significantly contribute to an inclusive transport system, market-driven delivery of such services might render them too costly for a substantial portion of the underserved population, raising equity concerns. Additionally, such services are likely to be concentrated in high-demand areas, which often already offer high levels of mobility and accessibility through mixed land use patterns in combination with by cycling infrastructure and a dense public transport network. Benefits could be more widely spread if such sharing services were regulated to guarantee inclusive design and subsidized to guarantee availability and affordability to a wider share of the population.

Our analysis indicates that the greatest environmental and equity benefits are likely when ACT technologies are employed to enhance public transport. Broader utilization of driver support technologies across all public transport modes may result in modest cost reductions and improvements in passenger comfort (Cohen & Hopkins, 2019; Thomopoulos et al., 2021). Full automation may substantially further increase the availability and quality of public transport, in terms of frequencies, operating hours, and coverage area, as well as convenience and comfort, through a combination of schedule-based and complimentary on-demand services. Some of these improvements are possible even within current public transport budgets, which may lead to an increase in ridership and subsequently to renewed policy attention and public budgets for further improvements in public transport service, with likely positive effects on ridership and revenues and related social and environmental sustainability benefits. Given that the other ACT deployment types present significant use barriers for a range of population groups, we conclude that ACT-based public transport offers the most potential toward an inclusive transport system. While this holds for developed car-oriented societies, it is also the most promising direction for emerging economies and countries of the Global South, where the vast majority of the population is unlikely to have access to a private (ACT) vehicle in the foreseeable future.

The kind of ACT-based future transport system that will emerge will depend not so much on technology but on policy. The case of Singapore underscores the potential role of governments in leveraging ACT-based technologies toward a more inclusive transport system. The Singapore experience is comparable to other experiments worldwide, showing that fully automated public transport is still some years away. Yet, Singapore stands out in two respects. First, it is moving toward a comprehensive ACT ecosystem, which provides a rich infrastructure for ongoing experimentation with ACT-based (public) transport. Second, and more importantly, it is committed to employing this ecosystem for the greater public good rather than merely for private profit.

Singapore is experimenting with a range of technologies, such as precision docking, platooning and real-time fleet management, all of which may improve the quality of public transport services. While these are small steps, they can contribute to a more inclusive transport system by improving the cost balance of public transport. Small improvements may attract new riders and thus increase revenues, as well as reduce operational costs. If these revenues and reduced costs are funneled back into the public transport system, modest ACT may well assist in creating a virtuous cycle and thus contribute to a more inclusive transport system.

ACKNOWLEDGMENTS

This chapter is in part based on the extensive discussions within Working Group 2 (WG2) of WISE-ACT (EU COST Action CA16222). A more elaborate discussion of the relationship between ACT and inclusion is available in Martens et al. (2021). This chapter has especially benefitted from the inputs from WG2 members Bert van Wee, Milos Mladenovic, and Dimitrios Milakis, as well as from the editors of this book. We also want to thank Einav Henenson for her research support. The responsibility for any mistakes in the chapter remains solely with the authors.

REFERENCES

Allen, J., & Farber, S. (2021). Suburbanization of transport poverty. *Annals of the American Association of Geographers*, *111*(6), 1833–1850.

Banister, D. (1999). Planning more to travel less: Land use and transport. *Town Planning Review (1999)*, *70*(3), 313–338.

Beyazit, E., & Sungur, C. (2019). Working women and unequal mobilities in the urban periphery. In R. Hickman, B. M. Lira, M. Givoni, & K. Geurs (Eds.), *A companion to transport, space and equity* (Chapter 11, pp. 147–166). Edward Elgar Publishing.

Blyth, P. L., Mladenovic, M. N., Nardi, B. A., Ekbia, H. R., & Su, N. M. (2016). Expanding the design horizon for self-driving vehicles: Distributing benefits and burdens. *IEEE Technology and Society Magazine*, *35*(3), 44–49.

Bösch, P. M., Becker, F., Becker, H., & Axhausen, K. W. (2018). Cost-based analysis of autonomous mobility services. *Transport Policy*, *64*, 76–91.

Chong, Z. J., Qin, B., Bandyopadhyay, T., Wongpiromsarn, T., Rankin, E. S., & Ang, M. H. (2011). Autonomous personal vehicle for the first- and last-mile transportation services. In *IEEE 5th International Conference on Cybernetics and Intelligent Systems (CIS)*, pp. 253–260. https://doi.org/10.1109/ICCIS.2011.6070337

Chng, S., Kong, P., Lim, P. Y., Cornet, H., & Cheah, L. (2021). Engaging citizens in driverless mobility: Insights from a global dialogue for research, design and policy. *Transportation Research Interdisciplinary Perspectives*, *11*, 1–14. https://doi.org/10.1016/j.trip.2021.100443

Clark, J., & Curl, A. (2016). Bicycle and car share schemes as inclusive modes of travel? A socio-spatial analysis in Glasgow. *Social Inclusion*, *4*(3), 83–99.

Cohen, T., & Cavoli, C. (2019). Automated vehicles: Exploring possible consequences of government (non)intervention for congestion and accessibility. *Transport Reviews*, *39*, 129–151.

Cohen, S. A. & Hopkins, D. (2019). Autonomous vehicles and the future of urban tourism. *Annals of Tourism Research*, *74*, 33–42.

Cornet, H., Stadler, S., Kong, P., Marinkovic, G., Sathikh, P. M, & Frenkler, F. (2019). User-centred design of autonomous mobility for public transportation in Singapore. *Transportation Research Procedia*, *41*, 191–203.

Costantini, F., Thomopoulos, N., Steibel, F., Curl, Kovacikova, T., & Lugano, G. (2020). Autonomous vehicles in a GDPR era: An international comparison. In D. Milakis, N. Thomopoulos, & B. van Wee (Eds.), *Policy implications of autonomous vehicles*. Academic Press.

Cugurullo, F., & Acheampong, R. A. (2023). Fear of AI: An inquiry into the adoption of autonomous cars in spite of fear, and a theoretical framework for the study of artificial intelligence technology acceptance. *AI & Soc*. https://doi.org/10.1007/s00146-022-01598-6

Curl, A., & Fitt, H. (2018). *What do driverless cars mean for cities, health and wellbeing? Proceedings of the New Zealand Geographical Society (NZGS) and the Institute of Australian Geographers (IAG) Conference*, p. 30. https://www.iag.org.au/client_images/2092803.pdf

Currie, G., & Senbergs, Z. (2007). *Exploring forced car ownership in metropolitan Melbourne*. Australian Transport Research forum 2007. https://www.researchgate.net/profile/Graham-Currie/publication/37183729_Exploring_forced_car_ownership_in_metropolitan_Melbourne/links/0fcfd5097adf85a8d2000000/Exploring-forced-car-ownership-in-metropolitan-Melbourne.pdf

De Madariaga, S. (2013). Mobility of care: Introducing new concepts in urban transport. In S. De Madariaga & M. Roberts (Eds.), *Fair shared cities*. Routledge.

Dellinger, A. M., Sehgal, M., Sleet, D. A., & Barrett-Connor, E. (2001). Driving cessation: What older former drivers tell us. *Journal of the American Geriatrics Society*, *49*(4), 431–443.

Docherty, I., Stone, J., Curtis, C., Sørensen, C. H., Paulsson, A., Legacy, C., & Marsden, G. (2022). The case for "public" transport in the age of automated mobility. *Cities*, *128*, 103784. https://doi.org/10.1016/j.cities.2022.103784

Etminani-Ghasrodashti, R., Ketankumar Patel, R., Kermanshachi, S., Michael Rosenberger, J., & Weinreich, D. (2021). Exploring concerns and preferences towards using autonomous vehicles as a public transportation option: Perspectives from a public focus group study. In *International conference on transportation and development 2021*, 344–354. https://ascelibrary.org/doi/abs/10.1061/9780784483534.030

Forth Mobility. (2020). *Low-income carsharing report*. https://forthmobility.org/storage/app/media/uploaded-files/Low-Income_Car-Sharing_Report_3.16.pdf

Groshen, E. L., Helper, S., MacDuffie, J. P., & Carson, C. (2019). *Preparing U.S. workers and employers for an autonomous vehicle future*. [Upjohn Institute Technical Report No. 19-036. W.E. Upjohn Institute for Employment Research].

Gwyther, H., & Holland, C. (2012). The effect of age, gender and attitudes on self-regulation in driving. *Accident Analysis and Prevention*, *45*, 19–28.

Guo, J., Susilo, Y., Antoniou, C., & Pernestal, A. (2020). Influence of individual perceptions on the decision to adopt automated bus services. *Sustainability*, *12*(16), 6484.

Guo, J., Susilo, Y., Antoniou, C., & Pernestal, A. (2021). When and why do people choose automated buses over conventional buses? Results of a context-dependent stated choice experiment. *Sustainable Cities and Society*, *69*, 102842.

Harper, C. D., Hendrickson, C. T., Mangones, S., & Samaras, C. (2016). Estimating potential increases in travel with autonomous vehicles for the non-driving, elderly and people with travel-restrictive medical conditions. *Transportation Research Part C: Emerging Technologies*, *72*, 1–9.

Herzogenrath-Amelung, H., Troullinou, P., & Thomopoulos, N. (2015). Reversing the order: Towards a philosophically informed debate on ICT for transport. In N. Thomopoulos, M. Givoni, & P. Rietveld (Eds.), *ICT for transport: Opportunities and threats*. Edward Elgar.

Horizon 2020 Commission Expert Group. (2020). To advise on specific ethical issues raised by driverless mobility (E03659). *Ethics of Connected and Automated Vehicles: recommendations on road safety, privacy, fairness, explainability and responsibility*. Publication Office of the European Union: Luxembourg.

Kavanagh, A. M., Krnjacki, L., Aitken, Z., Lamontagne, A. D., Beer, A., Baker, E., & Bentley, R. (2015). Intersections between disability, type of impairment, gender and socio-economic disadvantage in a nationally representative sample of 33,101 working-aged Australians. *Disability and Health Journal*, *8*(2), 191–199.

Kim, K. (2015). Can carsharing meet the mobility needs for the low-income neighborhoods? Lessons from carsharing usage patterns in New York City. *Transportation Research Part A: Policy and Practice*, *77*, 249–260.

Koppel, S., Lee, Y.-C., Mirman, J. H., Peiris, S., & Tremoulet, P. (2021). Key factors associated with Australian parents' willingness to use an automated vehicle to transport their unaccompanied children. *Transportation Research Part F: Traffic Psychology and Behaviour*, *78*, 137–152.

Knupfer, S. M., Pokotilo, V., & Woetzel, J. (2018). *Elements of success: Urban transportation systems of 24 global cities*. McKinsey & Company.

Kumar Mitra, S. (2021). Impact of carsharing on the mobility of lower-income populations in California. *Travel Behaviour and Society*, *24*, 81–94.

Kyriakidis, M., Sodnic, J., Stojmenova, K., Elvarsson, A., Pronello, C., & Thomopoulos, N. (2020). The role of human operators in safety perception of AV deployment. *Sustainability*, *12*(21), 9166.

Kurniawan, J. H., Chng, S., & Cheah, L. (2021). The social acceptance of autonomous vehicles, *IEEE Potentials*, *40*(4), 39–44.

Land Transport Authority (LTA). (2016, October 19). *Joint News Release by the Land Transport Authority (LTA) and NTU – On the Road to a More Sustainable and Reliable Transport System,*

News Release, Land Transport Authority. https://www.lta.gov.sg/content/ltagov/en/newsroom/2016/10/2/joint-news-release-by-the-land-transport-authority-lta-ntu--on-the-road-to-a-more-sustainable-and-reliable-transport-s.html

Land Transport Authority (LTA). (2017, April 10). *LTA Inks Agreement with ST Kinetics to Develop and Trial Autonomous Buses, News Release, Land Transport Authority.* https://www.lta.gov.sg/content/ltagov/en/newsroom/2017/4/2/lta-inks-agreement-with-st-kinetics-to-develop-and-trial-autonomous-buses.html

Land Transport Authority (LTA). (2019). *Land Transport Master Plan 2040, Land Transport Authority (LTA), Singapore.* https://www.lta.gov.sg/content/ltagov/en/who_we_are/our_work/land_transport_master_plan_2040.html

Land Transport Authority (LTA). (2021, September 3). *Joint Factsheet by the Land Transport Authority (LTA) and SSC – Enhanced National Standards for the Safe Deployment of Autonomous Vehicles in Singapore.* https://www.lta.gov.sg/content/ltagov/en/newsroom/2021/9/news-releases/enhanced-national-standards-for-the-safe-deployment-of-autonomou.html

Land Transport Authority (LTA). (n.d.). *LTA website, Autonomous Vehicles, Land Transport Authority (LTA), Singapore.* https://www.lta.gov.sg/content/ltagov/en/industry_innovations/technologies/autonomous_vehicles.html

Levinson, D. (2018). *Road rent – On the opportunity cost of land used for roads* [web log posted]. Retrieved December 21, 2018, from https://transportist.org/category/parking

Litman, T. (2020). *Evaluating household chauffeuring burdens.* Victoria Transport Policy Institute. https://www.vtpi.org/chauffeuring.pdf

Lim, P. Y., Kong, P., Cornet, H., & Frenkler, F. (2021). Facilitating independent commuting among individuals with autism – A design study in Singapore. *Journal of Transport and Health, 21,* 101022.

López, C., Ruíz-Benítez, R., & Vargas-Machuca, C. (2019). On the environmental and social sustainability of technological innovations in urban bus transport: The EU Case. *Sustainability, 11*(5), 1413.

Lucas, K. (2012). Transport and social exclusion: Where are we now? *Transport Policy, 20,* 105–113.

Marsden, G., Anable, J., Bray, J., Seagriff, E., & Spurling, N. (2019). *Shared mobility: Where now? Where next? The second report of the Commission on Travel Demand.* Centre for Research into Energy Demand Solutions.

Martens, K. (2017). *Transport justice: Designing fair transportation systems.* Routledge.

Martens, K. (2018). Ageing, impairments and travel: Priority setting for an inclusive transport system. *Transport Policy, 63,* 122–130.

Martens, K., & Lucas, K. (2018). Perspectives on transport and social justice. In G. Craig (Ed.), *Handbook on global social justice* (pp. 351–370). Edward Elgar.

Martens, K., Beyazit, E., Henenson, E., Thomopoulos, N., Milakis, D., Mladenovic, M., Pudane, B., van Wee, B., Di Ciommo, F., Curl, A., Cugurullo, F., Dimitrova, E., & Negulescu, M. (2021). *WG2 Thematic Report: Autonomous and connected transport as part of an inclusive transport system,* COST Action CA16222 WISE-ACT.

Martin, G. (2019). *Sustainability prospects for automated vehicles: Environmental, social and urban.* Routledge.

Mattioli, G. (2017). 'Forced car ownership' in the UK and Germany: Socio-spatial patterns and potential economic stress impacts. *Social Inclusion, 5*(4), 147–160.

Mattioli, G., Lucas, K., & Marsden, G. (2017). Transport poverty and fuel poverty in the UK: From analogy to comparison. *Transport Policy, 59,* 93–105.

Milakis, D., & van Wee, B. (2020). Implications of vehicle automation for accessibility and social inclusion of people on low income, people with physical and sensory disabilities and older people. In C. Antoniou, D. Efthymiou, & E. Chaniotakis (Eds.), *Demand for emerging transportation systems. Modeling adoption, satisfaction and mobility patterns* (pp. 61–73). Elsevier.

Mo, B., Cao, Z., Zhang, H., Shen, Y., & Zhao, J. (2021). Competition between shared autonomous vehicles and public transit: A case study in Singapore. *Transportation Research Part C-Emerging Technologies, 127,* 103058.

Mobile Lives Forum. (2021). *Autonomous vehicles: What role do they have in the mobility transition?* https://en.forumviesmobiles.org/project/2020/02/07/autonomous-vehicles-what-role-do-they-have-mobility-transition-13212?utm_source=nl_en&utm_medium=email&utm_campaign=0903_AutoVeh

Newman, P., & Kenworthy, J. (1999). *Sustainability and cities: Overcoming automobile dependence.* Island Press.

Nikitas, A., Thomopoulos, N., & Milakis, D. (2021). The environmental and resource dimensions of automated transport: A nexus for enabling vehicle automation to support sustainable urban mobility. *Annual Review of Environment and Resources*, *46*, 167–192.

Ongel, A., Cornet, H., Kong, P., Khoo, R., Liu, T., & Kloeppel, M. (2018). Public transport service quality improvement using universal design standards and advanced vehicle technologies. In *International Conference on Intelligent Autonomous Systems (ICoIAS)*, pp. 211–216.

Park, J., & Chowdhury, S. (2018). Investigating the barriers in a typical journey by public transport users with disabilities. *Journal of Transport and Health*, *10*, 361–368.

Parker, D., Macdonald, L., Sutcliffe, P., & Rabbitt, P. (2001). Confidence and the older driver. *Ageing and Society*, *21*(2), 169–182.

Patel, R. K., Etminani-Ghasrodashti, R., Kermanshachi, S., Rosenberger, J. M., & Weinreich, D. (2021). Exploring preferences towards integrating the autonomous vehicles with the current microtransit services: A disability focus group study. In *International conference on transportation and development 2021*, pp. 355–366. https://ascelibrary.org/doi/abs/10.1061/9780784483534.031

Polydoropoulou, A., Tsouros, I., Thomopoulos, N., Pronello, C., Elvarsson, A., Sigþórsson, H., Dadashzadeh, N., Stojmenova, K., Sodnik, J., Neophytou, S., Esztergár-Kiss, D., Hamadneh, J., Parkhurst, G., Etzioni, S., Shiftan, Y., Di Ciommo, F. (2021). Who is willing to share their AV? Insights about gender differences among seven countries. *Sustainability*, *13*(9), 4769.

Rosenbloom, S. (2010). How adult children in the UK and the US view the driving cessation of their parents: Is a policy window opening? *Journal of Transport Geography*, *18*(5), 634–641.

Shaheen, S., Bell, C., Cohen, A., & Yelchuru, B. (2017). *Travel behavior: Shared mobility and transportation equity*. https://www.fhwa.dot.gov/policy/otps/shared_use_mobility_equity_final.pdf

Shaheen, S. A., Cohen, A. P., Broader, J., Davis, R., Brown, L., Neelakantan, R., & Gopalakrishna, D. (2020). *Mobility on demand planning and implementation: Current practices, innovations, and emerging mobility futures*. https://rosap.ntl.bts.gov/view/dot/50553

Shen, Y., Zhang, H., & Zhao, J. (2018). Integrating shared autonomous vehicles in public transportation systems: A supply-side simulation of the first-mile service in Singapore. *Transportation Research Part A: Policy and Practice*, *113*, 125–136.

Shiftan, Y., Polydoropoulou, A., Thomopoulos, N., & Rappazzo, V. (2021). Autonomous and connected transport – The user perspective. Introduction to the special issue. *Sustainability*, *13*(9). https://www.mdpi.com/journal/sustainability/special_issues/Automated_Connected_Transport

Sparrow, R., & Howard, M. (2017). When human beings are like drunk robots: Driverless vehicles, ethics, and the future of transport. *Transportation Research Part C: Emerging Technologies*, *80*, 206–215.

Soh, E., & Martens, K. (2023). Public values in the socio-technical construction of autonomous vehicle futures. *Public Management Review*, 1–19.

Society of Automotive Engineers International (SAE). (2021). *Taxonomy and definitions for terms related to driving automation systems for on-road motor vehicles*. Revised version 30 April 2021. https://www.sae.org/standards/content/j3016_202104/

Thomopoulos, N., Cohen, S., Hopkins, D., Siegel, L., & Kimber, S. (2021). All work and no play? Autonomous vehicles and non-commuting journeys. *Transport Reviews*, *41*(4), 456–477.

Thomopoulos, N., & Givoni, N. (2015). The autonomous car: A blessing or a curse for the future of low carbon mobility? An exploration of likely vs. desirable outcomes. *European Journal of Futures Research*, *3*(14), 1–14.

Tice, P. C. (2019). Micromobility and the built environment. *Proceedings of the Human Factors and Ergonomics Society Annual Meeting*, *63*(1), 929–932.

Tirachini, A. (2020). The economics of automated public transport: Effects on operator cost, travel time, fare and subsidy. *Economics of Transportation*, *21*, 100151.

Uteng, T. P. (2009). Gender, ethnicity, and constrained mobility: Insights into the resultant social exclusion. *Environment and Planning A*, *41*, 1055–1071.

Wadud, Z., MacKenzie, D., & Leiby, P. (2016). Help or hindrance? The travel, energy and carbon impacts of highly automated vehicles. *Transportation Research Part A: Policy and Practice*, *86*, 1–18.

Young, M., & Farber, S. (2021). Ride-hailing platforms are shaping the future of mobility, but for whom? In A. Austin Zwick & Z. Spicer (Eds.), *The platform economy and the city: Urban peril and promise in the new digital economy*. McGill-Queens University Press.

CHAPTER 9

VIABLE BUSINESS OR VITAL ENVIRONMENT? DECONSTRUCTING THE SUSTAINABILITY CONCEPT IN FUTURE MOBILITY ENTREPRENEURSHIP

Graham Parkhurst[a], Pablo Cabanelas[b] and Daniela Paddeu[a]

[a]*Centre for Transport and Society, University of the West of England, Bristol, UK*
[b]*Faculty of Commerce, University of Vigo, Spain*

ABSTRACT

Rapid technological change in the transport sector is leading to a growing range of potential and actual 'business models' deployable for the movement of goods and people. Two key uncertainties arise from this proliferation: first, concerning which ones can be economically viable, and, second, whether they can be both simultaneously economically viable and contribute to the imperatives of more sustainable mobility. The present chapter reviews and appraises the emergence of these new business models, drawing on both literature review and empirical research with entrepreneurs involved in the new mobility sector. Specifically, the potential of the UN Sustainable Development Goals (SDGs)

Sustainable Automated and Connected Transport
Transport and Sustainability, Volume 19, 161–185

ISSN: 2044-9941/doi:10.1108/S2044-994120240000019009

(UN, n.d.) as a device to structure and frame the debate about what constitutes a valuable contribution to sustainable mobility is considered. A framework is developed which captures how mobility and transport have dependencies with the SDGs. From this analysis, key sustainability concepts are derived which have either a subsistence function (maintaining the basics of human life) or an enhancement function (enabling citizens to realise their potential whilst reducing impacts on the planet). Five different innovations involving mobility sector business entrepreneurship are then characterised using this framework to exemplify its ability to deconstruct and test claims that 'smart mobility' is also good for sustainability as well as good for business. It is concluded that the framework could contribute to a wider architecture of sustainability interrogation. It could promote discourse around a wide range of actors, posing questions and surfacing tensions and contingencies effectively, whilst providing a holistic, strategic assessment to inform more targeted, scientific evaluations of sustainability metrics.

Keywords: Sustainable development goals; sustainable mobility discourse; future mobility; green growth; stakeholder engagement framework; electric cars; hydrogen heavy goods vehicles; Mobility-as-a-Service; unmanned aerial vehicles; autonomous delivery robots

INTRODUCTION

The last decade has seen a striking level of entrepreneurial activity in the transport sector, covering both the passenger and freight markets. Commentators refer to the 'new', 'smart' or 'future' mobilities in indicating a significant change in the nature and extent of technologies deployable for the movement of goods and people. This technological revision, if not revolution, leaves many proponents without doubt as to the economic growth potential of the sector in aggregate terms (e.g. Bouton et al., 2015; Graham, 2013; Lerner, 2011; Van Audenhove et al., 2014). However, for the individual company engaged in the daily evolution of 'future mobility', this economic prize is mostly far from secure and obvious, due to major risks about technology choices and policy stability, let alone finding demand in the marketplace. Even giants such as Tesla and Uber have struggled to achieve profitability, despite huge stock-market valuations. And as well as future mobility representing an economic opportunity, the implications for the environmental and social problems that the transport sector faces are also potentially enormous. Indeed, it is fundamental that technological change to new 'business models' must be accompanied by making a significant difference in performance against these criteria, variously at the local and global levels.

The new technologies are for the most part underpinned by growing digitalisation, which provides new options for social and economic development through new business models (Caputo et al., 2021; Pagani & Pardo, 2017). For example, connecting vehicles with each other and infrastructure leads to new business ecosystems (Epting, 2018). Here, a central proposal of the new market offers in the area of mobility is the increasing efficiency in operations leading to lower climate change impact (Milakis et al., 2020; Nikitas et al., 2021; Parkhurst & Seedhouse, 2019). But this path towards more sustainable and social mobility has at best only just started. It is necessary to achieve the commitment of a very wide group of actors and stakeholders, within and beyond the transport sector, to mitigate the environmental impact of the projected growth both in freight and passenger demand. Estimates from prior to the COVID-19 pandemic were for 33% growth in freight and 16% growth in passenger traffic by 2030, and 55% and 30% by 2050 (JRC, 2019).

Original Equipment Manufacturers (OEM) are committed to the development of digital technologies and their application to mobility (Marletto, 2019), for example through automation and connectivity, variously 'Connected Autonomous Vehicles' or 'Connected Automated Vehicles' (CAV) (Shergold, 2019; Whittle et al., 2019). CAVs represent a major innovation associated with a disruptive technology, with the capacity to transform the current mobility ecosystem. The industry and its regulators are at a crossroads in deciding how to adapt the regulatory frameworks to environmental challenges whilst providing an adequate background for viable business models (Athanasopoulou et al., 2019; Rode et al., 2017; Sarasini & Linder, 2018). However, the uncertainty and the complexity of the new reality make it necessary for the establishment of new alliances to understand new technologies and address customer demands through interdisciplinary approaches that go beyond traditional players (Pütz et al., 2019). Therefore, the boundaries of the mobility value chain are experiencing important changes, at the same time that they are being blurred because new actors appear with supposedly better solutions to address new needs from markets (Oskam et al., 2021). Hence, in this time of change, whilst evolution ought to be directed to favour more socially inclusive and environmentally affordable mobility fostered by new technologies, the criteria of profitability may constitute a dilemma, or worse a barrier, to such positive change.

The chapter contributes to resolving this dilemma by providing an initial analysis of some of the key technologies and business models associated with the transition. It begins in the next section by identifying in more detail the principal 'disruptive' technologies in the freight and passenger sectors. It then turns in the third section to propose a framework to consider how the United Nations' SDGs can be used to frame the policy debate amongst diverse stakeholders about these emergent technologies. The following section then provides example applications of the framework, followed in the final section by consideration of the relevance of the approach for policy-makers seeking to assess innovations.

DISRUPTIVE FUTURE TECHNOLOGIES

Freight Mobility

Jaller and Pahwa (2020) argue that the key to future freight sustainability is a combination of automation, electrification and shared resources in order to improve efficiency and effectiveness. Road transport accounts for 76.3% of freight movements in Europe (Eurostat, 2021), and it is responsible for more than 40% of global carbon emissions from transport, with road freight alone accounting for 29.4% (IEA, 2022). A series of technological solutions will be available in the medium to long term to improve the sustainability of freight movements. The use of hydrogen and electricity as energy vectors is expected to significantly impact emissions reduction during longer distance movements (Greening et al., 2019). However, there is deep uncertainty towards future operations (e.g. readiness of technology, infrastructure requirements) and costs (e.g. production of fuels and vehicles, operational and maintenance). In particular, the future usage of hydrogen will depend on the cost and feasibility of its production technology. In fact, generation, storage, and movement of hydrogen reduce its efficiency in comparison to electricity and therefore might suggest it would be relatively uncompetitive (Greening et al., 2019). However, hydrogen remains within the policy-technology frame due to the other main option, electricity, also having significant problems with commercial applications.

Electrification of all modes of freight is seen as a potential option in the long term (Parkhurst, 2021). However, numerous challenges to commercial application exist related to feasibility, especially in terms of supply (e.g. availability of electricity to power the whole fleet), network design (e.g. recharging stations), technological development (especially for heavier road freight vehicles, ships, and aircraft) and overall costs. The potential demands on battery technology imply large size and weight, and so are expected to significantly reduce the vehicle's carrying capacity. Reduced capacity relative to other power trains threatens the financial viability of battery–electric business models, even if hitherto there is a lack of real-world validation of anticipated efficiencies and weight/payload availabilities (Haugen et al., 2021). Some concepts identify business models involving radical technology transfer, including hence high infrastructure costs and integration challenges; for example, using pantographs common in the rail sector to create 'electric highways' with overhead power supply by contact with powered wires. Heavy goods vehicles (HGVs) using such a system can be lighter due to the use of a smaller battery, and consequently can carry a greater payload, and can therefore perform better in terms of value for money (Siemens, 2017). In addition, even though electric batteries would not be an option for most shipping business models, Smith et al. (2019) suggest that there are niches in which batteries can be used to power ships that travel short distances, navigating within and between ports, so able to refuel regularly.

In addition to cleaner future technologies, there might be the opportunity to use automation to increase efficiency, whilst improving the sustainability of freight movements (Paddeu & Parkhurst, 2020). These can include autonomous vessels, aircraft and HGVs on the long haul, and autonomous delivery vehicles, including autonomous ground pods and robots, and unmanned aerial vehicles (e.g. drones) (Paddeu et al., 2019). In particular, PATH (2017) and MTG (2019) estimated that there would be a potential for up to €6 billion a year in fuel cost savings within Europe. However, there is deep uncertainty about the actual rate of fuel savings, as the technology is still under development (Tavasszy, 2016), and will significantly depend on the design of the road infrastructure and traffic conditions (Bakermans, 2016; Paddeu & Denby, 2021). Also, the investment cost of substituting old for new technologies is often unaffordable for small and medium enterprises, and, for all businesses, the improved efficiency rate is not yet sufficient to justify investment (Wiegmans et al., 2007). Nonetheless, if a viable business model emerges, the lack of access to capital or lack of economy of scale may be a cause of inequality in the future freight market, with bigger companies gaining an advantage over smaller operators. In addition, the creation of new services for digitalisation and automation will encourage new players to enter the market (Paddeu et al., 2019).

Passenger Mobility

Passenger mobility business models are seen to be at a 'crossroads' due also to societal trends as well as technological evolution, or perhaps more properly due to a *coevolution* of those factors. Recent social generations (namely Generation Y 'the millennials' and Generation Z 'the digital natives') are identified as being less interested in owning vehicles (Focas & Christidis, 2017; Turienzo et al., 2022a) and possessing a driving licence (Tilley & Houston, 2016). At the same time, they are found to be more interested in mobility options identified as more sustainable, like walking, bus use, and cycling, among others (Whittle et al., 2019). In part, this is thought to arise through greater awareness of climate change, but also due to increased collaborative and shared transport encouraged by regulators (Eckhardt et al., 2018; European Commission, 2018). Here, 'shared' can mean both synchronous or asynchronous use of transport assets by different parties (Parkhurst & Seedhouse, 2019).

This co-evolution opens an opportunity window to develop business models for more collaborative and shared transport niches. Those new technologies favour contact and exchange among people through technological platforms that permit a higher personalisation of services (Ahmed et al., 2020; Athanasopoulou et al., 2019). One such example is subscription business models for packages identified as 'Mobility-as-a-Service' (Jittrapirom et al., 2018). Providing collaborative services requires a change in the mindsets of participating businesses, which will need to take part in alliances of businesses, adopt a multi-disciplinary perspective (Pütz et al., 2019) and ultimately be willing to 'share' customers. Hence, the mobility value chain is increasingly blurring the boundaries as new participants address new market demands, particularly technological companies (Oskam et al., 2021). Some of these new business model niches are identified by their proponents as seeking to provide cleaner, lower carbon emission and more efficient mobility solutions featuring platforms (Riggs & Beiker, 2019). One such example is the consolidation of demand and supply through dynamic simulations (Berg et al., 2020) in order to predict where demand might arise. Another is the negotiation of asset sharing, as in the case of shared-ride on-demand services (Calvert et al., 2019), facilitated by the integration of connected vehicles in the system (Turienzo et al., 2022a).

Such changes mean a new form of competition for traditional OEMs (such as Volkswagen or Toyota) but also for all mobility-related business models, including mobility support services (e.g. maintenance workshops, highway service area operators and insurance companies). Both traditional and newcomer players report adopting demanding sustainability principles and also to be adapting their value propositions to take advantage of technological possibilities. These responses include leveraging new digital platforms and providing a higher servitisation of their activities in order to offer more personalised mobility solutions (Cabanelas et al., 2023). Examples of sustainability commitments include a desire to use existing resources more efficiently, increasing the use and lifetime of vehicles and applying a circular economy logic (Alaerts et al., 2019). A practical demonstration of these approaches could include the better coordination of car-sharing fleets and the integration of electric or other lower environmental impact vehicles in the fleet (Globisch et al., 2019; Whittle et al., 2019).

Joining these initiatives is the principle that, if information flows between travellers and companies in a more connected way, the collective value proposition will be enhanced, and be directed towards increasing the sustainability of the whole passenger transport system. At the same time, better information can underpin greater individuality of service offers, and this could potentially increase the inclusivity of the system of mobility.

Whilst the rhetoric is clear and strong, whether these freight and passenger promises will be achieved must be subject to continual validation and analysis. The next section develops a framework for aligning stakeholder perspectives towards what constitutes a significant and deliverable shift towards sustainable mobility.

THE SUSTAINABLE DEVELOPMENT GOALS AS A FRAMEWORK FOR SUSTAINABLE MOBILITY POLICY DEVELOPMENT

The United Nations (UN) 17 Sustainable Development Goals (SDGs) (UN, n.d.) have been under development since the Rio de Janeiro Earth Summit of 1992 and are the centrepiece of the 2030 Agenda for Sustainable Development, adopted by UN Member States in 2015. The SDGs emphasise that environmental recovery and protection cannot, effectively or morally, proceed without integrated and concerted action to address social objectives including poverty, inequality and access to healthcare and education. Achieving these objectives will also require economic growth of a sympathetic nature.

The SDGs as a Tool for Sharing Perspectives on Sustainability

The SDGs represent a significant achievement in that they identify universal goals accepted by 193 national governments. However, they have been subject to a number of criticisms, summarised by ICSU/ISSC (2015). The SDGs can be seen as establishing endpoints without identifying process means of reaching them. Indeed, as an evaluation tool, the SDGs have been seen to be limited by the lack of quantifiable targets and timelines. The SDGs are also argued to be oriented towards the perspectives of national governments, rather than the wide range of stakeholders that need to be engaged to deliver them, notably those in the private sector. Furthermore, the 17 SDGs and their 169 targets had been identified as tending to promote 'silo' policy thinking, whereas there are important interactions between them and recognition, including by the UN, that they need to be overlaid by principles or themes which are even more fundamental, such as the 'six elements': people, planet, prosperity, justice, partnership, and dignity (UN, 2014).

Despite these criticisms, the SDGs nonetheless have a salience far beyond national governments and the public sector. Business consultants such as Ernst and Young extol the virtues of companies building them into strategic plans (Ernst & Young Global Ltd., 2017), and a survey has identified that more than four-fifths of companies use the SDGs as a framework against which to align their sustainability reports, even if they are limited as a tool for assessing progress (GRI, 2022). And whilst needing further development, the SDGs are also already a recognised 'brand' being applied at a wide scale in assessing company performance. The World Economic Forum (2021) reports on an assessment of 8,550 companies in the MSCI All Country World Index (finding that only a fraction were 'strongly aligned' and around a fifth 'aligned' with the goals).

Given their high recognition beyond the scientific community, and use by politicians, educators and business, the present chapter takes the view that the SDGs are, *de facto*, the framework within which different stakeholders are most likely to develop shared perspectives on development, including technological transitions. To this end, the chapter seeks to demonstrate how the SDGs can be applied to a specific case in order to enable the structuring of the complex reality of policy decisions (Schön & Rein, 1994). In this context, scientific predictive scenarios and ex-post evaluations are critical sources of information and evidence to test and validate the policy framings around the SDGs, but that framing is itself a distinct process of building (or not) a consensus around change. Indeed, as Genus and Stirling (2018) emphasise, exploring 'dissensus' is fundamental to questioning strongly asserted powerful interests and avoiding path-dependent or incumbent patterns of decision-making.

The approach also seeks to address the critiques that the SDGs must be applied in a relevant way and consider interdependencies. To this end, it explores an approach to grouping the SDGs applied to the case of transport and mobility policy. In doing so, it emphasises key themes in the delivery of change.

SDG Dependencies on Transport and Mobility

There is, perhaps rightly, no SDG for mobility or accessibility. The demand for transport services is mostly a 'derived demand', in that most of its socio-economic welfare benefits derive not from the movement itself, but in facilitating access to goods, services, restorative environments and social opportunities. However, for this very reason, several of the SDGs are contingent upon effective levels of physical accessibility. Depending on the socio-technical and built environment context, achieving accessibility often means using mobility services secured through one's own ownership of assets (e.g. car, bicycle) or renting assets owned by other (e.g. car hire) or paying to receive a transport service (e.g. bus, train or aeroplane journey). Table 9.1 summarises the principal dependencies between the SDGs and mobility in terms of the key motivations for movement (or, increasingly, and characterised strongly by practices during the COVID-19 pandemic, its substitution through virtual mobility).

Table 9.1 confirms that mobility policies and practices have implications for all 17 SDGs. In some cases (e.g. 9, 12, and 13), transport activity has a direct impact on their deliverability. In many, transport plays a key facilitating role (e.g. 2, 3, 4, and 8). For water (6) and energy (7), the preferred outcome is promoted by the extent people and commodities no longer have to travel, except by pipeline or wire. For some, transport and mobility present one important facet of the problem or development need (5, 10, 11, 15, and 16). Lastly, 17 represents the sector's ability to deliver change.

Taken at scale and in its full intent, the project of 'new mobility' identified in the introduction might have significant impacts on the broad dependencies within Table 9.1. Individual new mobility services will affect those relationships in specific ways, potentially with positive influence on some SDGs, and negative on others. Hence, the authors draw upon existing qualitative datasets developed through earlier projects which explored the emergence of new mobility services to examine those possible implications from the perspective of their architects and promotors. The data arise from individual expert interviews conducted in Spain, Portugal, and the United Kingdom with professionals involved in technology development in the transport sector, mainly focusing on passenger transport (MoBAE Project, CAPRI Project) and focus groups and workshops with professionals involved in the process of freight decarbonisation in the United Kingdom (CoDeZero and CRAFTeD projects). In each case, the data and results have been re-analysed through the lens of the SDGs.

For these analyses, the SDG dependencies are categorised into two groups (with SDG17 related to delivery outside those groups). The first group focuses on the connectivity that movement provides and how this directly underpins the SDGs which underpin subsistence in human life. The second has a more indirect, process-oriented relationship, which considers how the quality of transport and travel services influence the quality of life (Table 9.2).

APPLYING THE SDG-MOBILITY ANALYSIS FRAMEWORK

Taking the framework developed in the previous section, we apply it to five cases of future freight transport and passenger mobility technologies and their related business models. The cases variously consider freight and passenger mobility, air and surface transport systems, small-scale vehicles up to the largest road vehicles, and two cases which emphasise technology substitution, whereas three involve much more complex socio-technical and behavioural change. Two of the latter would use either automation or remote-control operation, for the third, automation could feature in the business model (Fig. 9.1).

Table 9.1. Principal Dependencies Between the SDGs and Mobility.

SDG Number and Scope	Dependencies on Mobility and Transport Services
1. End poverty in all its forms everywhere	Affordable access to a level of mobility services that facilitates at least the basic needs of life
2. End hunger, achieve food security and improved nutrition and promote sustainable agriculture	Efficient transport from the places of food production to markets via the transport system Physical or virtual access to marketplaces for food purchase, or via delivery services
3. Ensure healthy lives and promote well-being for all at all ages	Access to health services by physical means, although telemedicine of growing relevance Access to spaces which engender wellbeing (nature, historic environments, places of value to the individual), mainly physically, although virtual access is of relevance in some circumstances
4. Ensure inclusive and equitable quality education and promote lifelong learning opportunities for all	Physical and virtual access to the places of education and training – both modalities are now important due to the digitisation of knowledge and learning
5. Achieve gender equality and empower all women and girls	A transport and telecommunications system which offers services to all citizens without the threat or reality of physical or psychological harm during mobility, and which overcomes historic gender injustices in access to opportunity
6. Ensure availability and sustainable management of water and sanitation for all	Although for many citizens globally, high-quality water is found within the home or nearby, for many, travel for and transport of water is still part of the daily grind of chores. Enhanced water infrastructure is essential to minimise this need for travel
7. Ensure access to affordable, reliable, sustainable and modern energy for all	The transport of fossil fuels perpetuates global warming, adds to energy costs, and results in injury and deaths through fuel tanker collisions. Enhanced energy infrastructure is essential to bring clean, renewable energy to the home to minimise this need for transport
8. Promote sustained, inclusive and sustainable economic growth, full and productive employment and decent work for all	Physical mobility – daily or seasonal – to access a labour market with a range of opportunities is one of the traditional objectives of transport policy and planning. Increasingly, virtual access for remote working is also important to deliver access to employment

(*continues*)

Table 9.1. (*Continued*)

SDG Number and Scope	Dependencies on Mobility and Transport Services
9. Build resilient infrastructure, promote inclusive and sustainable industrialisation and foster innovation	Transport and communications services which use infrastructures in an efficient way, without imposing undue pressures on operating costs, capacity and longevity through extreme congestion or vehicle weight
10. Reduce inequality within and among countries	A level of international mobility of people and goods which enables resources to flow to places suffering inequality and enables people to migrate to places of opportunity without unsustainable levels of environmental impact
11. Make cities and human settlements inclusive, safe, resilient, and sustainable	Transport systems which serve their cities rather than dominating them to the exclusion of their primary purposes of being living spaces
12. Ensure sustainable consumption and production patterns	Linking the places of production and consumption with an intensity and efficiency that minimises energy demand and preserves local specialism
13. Take urgent action to combat climate change and its impacts	A transport system which will progressively and rapidly reach 'net zero'
14. Conserve and sustainably use the oceans, seas and marine resources for sustainable development	Marine transport systems with drastically reduced environmental pollution (combustion, waste and noise emissions) and physical disturbance
15. Protect, restore and promote sustainable use of terrestrial ecosystems, sustainably manage forests, combat desertification, and halt and reverse land degradation and halt biodiversity loss	Minimising the demand for additional land-take for transport infrastructure provision by using existing networks as efficiently as possible
16. Promote peaceful and inclusive societies for sustainable development, provide access to justice for all and build effective, accountable and inclusive institutions at all levels	Services are regulated to meet the needs of all citizens, so no one, literally and figuratively, is left behind
17. Strengthen the means of implementation and revitalize the global partnership for sustainable development	Ensuring that professionals, volunteers, and citizens engaged in sustainable mobility have the capacity and capability necessary to deliver against SDGs 1–16 as they relate to travel and transport locally, nationally and internationally

Table 9.2. SDG-Mobility Dependencies in terms of Subsistence and Enhancement of Quality of Life.

Group	Key Concepts from Dependencies
Outcome-related or subsistence SDGs	Movement to connect with goods and services
2/6/7 – hunger, thirst, comfort	• Food, water, energy
3 – health and wellbeing	• Healthcare, spaces for wellbeing
4 – education	• Education
8 – employment	• Employment
Process-related or enhancement SDGs	Quality of services for transport and travel
1 – poverty	• Affordability
5/10 – gender and other equality	• Inclusion: availability to all, security
9 – resilience	• Within technical system and ecosystem capacity
11 – urbanisation	• Human-scale infrastructure
12/14 – sustainability/conservation	• Energy efficient, low noxious emissions
13 – climate neutrality	• Low or zero GHG
15 – biodiversity	• Space efficient
16 – accountability	• Regulation in the interests of the public
Delivery-related SDG	Capacity and capability
17 – implementation	• Sufficient agents of change

Unmanned Aerial Vehicles (UAV)

Whilst passenger UAVs (pilotless aero taxis) are under development, freight drones are already offering some services. Drones can be used in combination with local transhipment hubs (such as consolidation centres) for 'last mile' deliveries. Possible applications include rural areas with low road transport accessibility and congested urban areas.

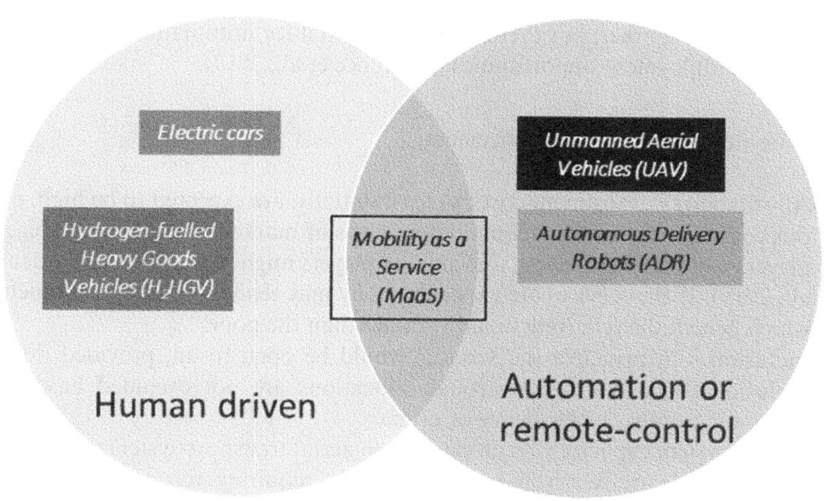

Fig. 9.1. Cases of Mobility Innovation With and Without Automation.

Drones can potentially support subsistence SDGs in the following ways, with providing alternatives to poor or absent infrastructure being a theme:

- Food – drones can be used to support smart agriculture (Kim et al., 2019), to increase production from smaller farms (Zhong et al., 2018) and as robot pollinators (Chechetka et al., 2017). A freight transport niche might be to transport high-value, low-weight produce to local markets in a timely fashion.
- Health and wellbeing – an established drone niche is for urgent deliveries of medical supplies to low-accessibility places, where the road infrastructure or congestion would not allow other efficient and timely road delivery options; or in humanitarian supply chains, such as emergencies and disasters (Singh & Frazier, 2018).
- Employment – they can be potentially used to replace vans in very congested areas (Paddeu et al., 2019). This might have a significant impact on employability: van drivers, for example, would be replaced, but new job profiles (drone handlers, drone operators-overseers), would be created. However, the literature does not provide any information on the specific impacts (Aurambout et al., 2019), but in some locations, drone services might be additional, so create employment, in others they may be an efficient alternative, requiring lower labour inputs. Drones also can be used for handling tasks in the local hub, as they have been shown to be significantly more efficient than human operators (Paddeu et al., 2019). For example, two drones can carry out the work of 100 humans (i.e., in terms of handling, picking and order preparation) over the same time period with an accuracy close to 100%, with warehousing and logistics cost savings (Jackson, 2017). These advantages might have a significant impact on the currently growing employment concerns within the logistics sector (Paddeu et al., 2019). The outcome would ultimately depend on the local labour market, but there is clear potential for automation in general to reduce employment opportunities (Lawrence et al., 2017).

Considering SDGs for enhancement:

- Affordability – investment costs, at least initially, are expected to be high, and this could tend to create inequalities in terms of market accessibility, as bigger players will be able to invest, and smaller players might be excluded (Paddeu et al., 2019). If the services are expensive, they may tend to be applied in niches which benefit the relatively wealthy, rather than the poor.
- Inclusion – in principle the services would be open to all, provided digital exclusion is avoided, and particular locations are not excluded based on economic viability or incidence of crime.
- Within system capacity – as drones are an aerial transport system, infrastructure demands are relatively low and specific, requiring vertical take-off and landing areas, or short runways. Along with all electric vehicles, there are

questions and uncertainties around the environmental impact of sourcing raw materials for batteries and their subsequent recycling.

- Human-scale infrastructure – hitherto most drones for civilian purposes have been small; indeed, fitting into the existing built environment is one advantage. This may not remain the case for all future applications, however.
- Energy and noxious emissions – aviation is fundamentally energy intensive but can represent the most efficient solution, e.g. if the alternative is an over-sized road vehicle taking an indirect route over hilly terrain. Noise emissions are a potential problem in urban areas and may limit operating practices.
- Low or zero GHG – assuming electric technology, there is the potential for zero-GHG emissions, depending on how electricity is produced (Portapas et al., 2021).
- Regulation in the interests of the public – probably the most significant barrier to the implementation of this technology is public acceptance. The public would need to be convinced that the services do not represent a threat to privacy, security or safety (Aurambout et al., 2019; Paddeu et al., 2019). There is a need for new regulations (Paddeu & Parkhurst, 2020).

Regarding capability and capacity (C&C), training would be needed not only for commercial operators but also for citizens in order for them to be able to interact with the technology (Paddeu et al., 2019).

Autonomous Delivery Robots (ADR)

Small-wheeled robots can be used to undertake B2C last-mile deliveries (and in this context, the typical range would be within a mile or so).

In respect of subsistence SDGs, ADRs have some similarities with the UAV equivalents, although they are constrained by using surface infrastructure and hence lower speeds due to interacting with other road users and obstacles in the built environment:

- Food – ADRs can be used for food last-mile deliveries. Whilst in many cases, this might be an added-value service to save the consumer a journey, some citizens may be unable to travel.
- Health and wellbeing – ADRs can deliver medical supplies, although in the case of time-critical deliveries may not offer the fastest option. During the COVID-19 pandemic, Abrar et al. (2020) suggested that ADRs could be a secure and contactless delivery option to counter the spread of the virus.
- Employment – similar to drones, ADRs could potentially replace human-driven vans in very congested areas, as they can circulate on the pavement sharing the space with pedestrians (Paddeu & Parkhurst, 2020). This might have a significant impact on employability (e.g. van drivers would be replaced). On the other hand, this could actually represent a partial solution to the truck/van driver shortage that is significantly impacting the logistics sector.

Considering SDGs for enhancement:

- Affordability – currently, the high initial investment costs of ADRs limit their widespread adoption. In fact, many companies that could potentially adopt ADRs can be expected to wait until the technology is further developed and tested (e.g. available and implemented on a large scale at a lower price), in order to avoid potential risks of misdirected investment (Zhang, 2019). Given the costs will need to be met by the consumer, until they reduce, the services are unlikely to be viable for most citizens.
- Inclusion – as with drones, there is a risk that some groups might be excluded, due to difficulties interacting with the digital and physical technologies (Paddeu et al., 2019), although ADRs could offer 24-hour delivery services, so meeting individual needs.
- Within system capacity – ADRs are designed to work with the existing infrastructure, although some adaptation may be necessary for their widespread adoption. All ADRs are likely to be electrically powered, adding to the transport system demand for batteries.
- Human-scale infrastructure – ADRs tend to be relatively small. However, they are intended for pavement use and therefore may constrain the use of pavements by humans wishing to exercise the most fundamental mobility right: walking.
- Energy and noxious emissions – as in the case of drones, energy efficiency will depend upon appropriate application, but there is potential to significantly reduce the number of failed deliveries (e.g. through offering evening delivery). Missed deliveries currently account for 5% of total e-commerce deliveries (Paddeu et al., 2019). Noxious emissions will be minimal due to electric power and low weight, so low tyre wear. Noise emissions will also be very low; a key advantage over drones.
- Low or zero GHG – being electric, ADR has the potential to reduce GHG emissions. However, it would be important to explore the added value they can provide in terms of environmental impact when compared with human-driven electric vehicles.
- Regulation in the interests of the public – at the moment, there are no regulations to allow ADRs to circulate on public roads. Previous studies identify resistance from pedestrians to sharing pavement space (Paddeu & Parkhurst, 2020), and it would be important to respect these views when seeking to update highway use codes to accommodate ADRs.

C&C is likely to be less safety-critical than an aerial system, but training for commercial operators and citizens will still be necessary (Paddeu et al., 2019).

Hydrogen-Fuelled Heavy Goods Vehicles (H2HGV)

The third case considering freight transport radically increases both the scale of mass being moved and the distance of typical operation (to medium-to-long distances). It also represents an example of technology substitution with minimal behaviour or context change required beyond the important exception of new fuel supply infrastructure provision.

For this reason, impacts on subsistence SDGs are likely to be relatively minor. The provision of goods to support food, health, education, etc. would be not directly affected by a switch to hydrogen, although access would be maintained whilst furthering other SDGs.

- Employment – the growth in the hydrogen industry has the potential to create a high number of new employment opportunities in a wide variety of industries (Bezdek, 2019). A recent study in the Netherlands estimates that the effect could be a net increase: green hydrogen could preserve and create up to 100,000 jobs a year by 2050, with one-third being new jobs created by the introduction of hydrogen (Energy Monitor, 2021).

Considering SDGs for enhancement:

- Affordability – companies wishing to buy hydrogen trucks would face high initial investment costs. Moreover, cost and feasibility remain very uncertain. The main rival, electrification, is not a viable option in the short to medium term, but in the long term, electric vehicles are expected to perform better both in terms of operations and costs (Greening et al., 2019). This increases the risk of investment for those companies who want to invest in hydrogen in the short term. Higher costs in the short run might be passed on to consumers, having a negative impact on poverty, although in the long run, the outcome might be positive, given the rising costs of fossil fuel extraction and production.
- Inclusion – alongside affordability and perceived risk, the lack of technical knowledge and infrastructures necessary for H_2 vehicles would, without intervention, be a barrier to participation in the transition, particularly affecting smaller operators.
- Within system capacity – the H_2 economy has the potential to underpin a circular economy. As in the case of electric powertrains, there are environmental impacts associated with the production of fuel cells, such as minerals used as catalysts, to be considered.
- Human-scale infrastructure – there is the potential for H_2 to be produced in a larger number of local facilities rather than at major refineries, which, if economically viable, would be at a more human scale.
- Energy and noxious emissions – practices for H_2 production and compression for use in road vehicle fuel cells mean that it is less energy efficient than the best electric technologies, but with advantages over internal combustion engines. Due to the only emissions from fuel cells being water, there are positive implications in terms of air quality and indirectly on public health due to the elimination of pollutants including oxides of nitrogen and particulates from exhausts. However, current production options can produce noxious by-products (Lewis, 2021).
- Low or zero GHG – depending on the way hydrogen is produced, HGV-H_2s have the potential to be zero GHG emissions (Greening et al., 2019).

C&C have a particular role within the freight sector to enhance the knowledge and skills necessary for the transition to H_2.

Mobility-as-a-Service (MaaS)

The *servitisation* of mobility involves the emergence of digital platforms to connect companies and travellers, and large amounts of data provided by connected vehicles. It therefore combines technological change with behavioural change, opening the prospect of significant system-wide changes which may facilitate, or obstruct, the achievement of the SDGs. However, the concept of MaaS covers a wide range of practical applications. Indeed, Mladenović and Haavisto (2021) identify the 'interpretive flexibility' of the concept as leading to a lack of clarity about the objectives and likely outcomes of the approach, creating a risk of unsubstantiated enthusiasm, even amongst public sector actors.

According to the nature of the implementation, the impact on SDGs can be expected to vary in important respects, and it would be important to interrogate the specific business model, not solely a generic concept. However, if new MaaS offers do facilitate travel behaviour change, particularly amongst those who suffer from accessibility constraints, the SDGs for Subsistence, in general, could potentially be strongly impacted, particularly with respect to:

- Health and wellbeing – access to services can be particularly problematic for disabled and older people where fixed transport schedules do not align with appointments. Access to on-demand transport with a wider range of options and alternatives (Shergold, 2019) can make personalised but collective transport affordable.
- Education – students might have more mobility opportunities, widening choice of school or higher education location and reducing constraints brought by limited accessibility (Turienzo et al., 2022a).
- Employment – labour market flexibility could be strongly enhanced (Calvert et al., 2022). However, some changes in employment roles in the transport sector can be expected due to the partial replacement of traditional public transport with MaaS. Further investigation is needed to better understand the impact of the digital economy, particularly on the need for low-skilled roles.

Considering SDGs for enhancement:

- Affordability – multimodal integration combined with a transfer from private to collective modes can reduce transport costs (Liu et al., 2017), offering more options to people with limited resources through a holistic design of pricing (Pangbourne et al., 2018). However, the extent to which access differences among social classes will persist likely depends on wider success in implementing the SDGs.
- Inclusion – if applied correctly, greater digitalisation will improve real-time information to identify and solve problems, to detect inappropriate behaviours

and to increase the accuracy of public information and extent of services, connecting the city better with the suburbs and rural areas. However, there remains a risk that personal data could be used inappropriately, to exclude particular citizens or social groups.

- Within system capacity – the high availability of behavioural data combined with the analytical power of companies and governments could facilitate scenario simulation to predict demand and the establishment of contingency processes to deal with unexpected phenomena. Such possibilities will be of particular benefit to road infrastructure managers (Turienzo et al., 2022a).
- Human-scale – will be improved by the higher efficiency and individuality of the transport system, using a greater share of micromobility (bikes, e-scooters, walking) rather than large vehicles.
- Energy and noxious emissions – the effective use of existing resources may enhance energy efficiency, but further evidence about the nature of behaviour change is necessary to evaluate whether, as a result of the service packages offered to consumers, the relatively efficient or inefficient modes benefit most. Emissions will depend on the powertrains of the selected options, but MaaS could promote the electrification of mobility, avoiding local exhaust emissions.
- Low or zero GHG – similarly, the integration of electric or other green technologies in the MaaS fleet would promote zero carbon emissions, but the fleet mix in practice needs to be better understood.
- Space efficiency – if transport assets such as vehicles and roads can be used more efficiently, this would reduce the pressure on natural resources (Alaerts et al., 2019).
- Regulation – a high level of coordination of the whole system will be necessary to facilitate the better use of existing resources and social inclusion. The high importance of personal data also raises protection and security matters. An important role for the state as a regulator will be necessary.

Significant C&C implications relate to the digitalisation of transport, requiring new skills in the sector, whilst the more traditional skills of integration will remain important.

Electric Cars (EC)

Here we consider the purely battery-powered plug-in EC, and no forms of hybrid, which, having dual powertrains, are more complex in respect of the SDGs. With several similarities to the case of HGV-H_2, the EC represents a 'business-largely-as-usual' approach to seeking greater sustainability for the private car. For this reason, the major change with respect to the subsistence SDG dependencies is with:

- Employment – even more so than in the case of HGV-H_2, there are major implications in terms of changing roles for vehicle assemblers and maintainers, and for the support services around charging rather than liquid fuelling.

Considering SDGs for enhancement:

- Affordability – in the short-run, ECs are like-for-like more expensive to produce than their internal combustion engine equivalents. However, those costs are expected to fall along with enhancements in battery technology, and there is a clear advantage in terms of operating costs with respect to fossil fuels for cars, even if all energy costs have been rising globally. In the long run, there is the potential for electric technology to lower the costs of transport products and services, offering some potential to reduce poverty (Börjesson et al., 2021).
- Inclusion – fundamentally, an EC is a car, and cars are associated with significant inequalities in mobility. New forms of service associated with the EC would be necessary to increase inclusion. Moreover, there is reason to believe that ECs could be a source of greater inequality, as some citizens who can own an ICE car would not be able to own an EC, due to charging constraints as well as capital costs (Parkhurst & Clayton, 2022).
- Within system capacity – reduced reliance on globally traded fossil fuels is expected to enhance the independence and resilience of some national economies, although dependence on specific commodities for battery production will likely create new dependencies. The new independence may be translated into public policies that foster the transition of their economies in a more balanced, competitive and socially equitable way. The EC support industry has the potential to offer a source of 'green jobs' and new business opportunities in the development of infrastructure and vehicles. The continuous technological evolution in this field is opening up promising areas of development. These could include start-ups for the development of new battery technologies and better battery management, including the circular economy around batteries (Turienzo et al., 2022b). Without an effective circular economy for batteries, there will be massive challenges for companies, policymakers and society in general as to how to source new raw materials and manage waste.
- Human scale – as the EC is essentially another kind of car, the human scale can only be respected by ensuring that alongside the transition to ECs, the personal-scale modes (micromobility) are given the greatest priority, otherwise our cities will continue to be congested with vehicles and the roads unsafe.
- Energy and noxious emissions – in many cities, ECs are subject to favourable regulatory and fiscal treatment, reflecting the benefits of zero tailpipe emissions and lower noise, expected to enhance the wellbeing and health of citizens (DBEIS, 2017). However, ECs are not generally being specified as energy-efficient vehicles, as they are designed to match or exceed the performance of the ICEs they replace, most of which are already over-specified for their 'duty cycle'. The energy-efficiency contribution of the transition to ECs would be enhanced by re-thinking the role of the car in society (Parkhurst & Clayton, 2022).
- Low or zero GHG – transport is responsible for almost 15% of greenhouse emissions worldwide, and road vehicles are the main source. Amongst those, cars are the largest contributor (Black et al., 2016; Greene & Parkhurst, 2017).

As noted under efficiency, ECs currently represent a partly missed opportunity. Moreover, to offer a benefit for all, the electrical energy must in the future arise from net-zero GHG sources; a major challenge for the electrical supply industry. Reduced demand for personal mobility by car may be the only means to overcome the challenge of sourcing sufficient green energy.

- Space efficiency – there are two areas of concern with ECs. First, by signalling greener and cheaper motoring to the consumer, demand for car travel may rise, and without new policies to manage demand, the need for additional roads, and hence loss of habitats, will grow. Second, whilst biodiversity may be favoured by reduced air pollution and GHG emissions, the challenges of raw material extraction for manufacturing batteries predominantly affect the industrialising states. However, reduced demand for new oil fields may reduce the risks faced by the few wild landscapes, such as those found in the Arctic or Antarctica.
- Regulation – the rise of the EC brings new challenges for regulators ranging from the road safety implications of their quietness compared to traditional cars, to who is responsible for organising an efficient network of recharging facilities. The state will also have an important role in managing the transition within an automotive sector that is traditionally a major provider of employment at a range of skill levels. The higher connectivity of future vehicles will also bring responsibilities to avoid cybercrime, whilst taking the opportunities for better highway and traffic management. Strong regulation of the circular economy around batteries will be necessary.

In respect of C&C, citizen-users of ECs will need only modest upskilling to change fuelling and driving practices, but major new skill sets are necessary for the sector to deliver charging infrastructure and to maintain the electric powertrain.

CONCLUSION: RELEVANCE OF THE FRAMEWORK FOR (DE)CONSTRUCTING SUSTAINABLE MOBILITY

The chapter could only examine five of the many business models and new technologies currently emerging in the sector. Nonetheless, the SDG-derived framework emerges as a relevant tool to encourage holistic thinking and debate around the impacts of new technology business models in the mobility sector. The approach reduces the 17 goals to a smaller number of mobility-oriented dependencies. Its particular value is a practical one of building upon the leading set of internationally endorsed sustainability criteria which are already in use by businesses and governments globally, and having high salience with citizens.

The specific assessments provided here are provisional, and no doubt partial. However, a notable concluding observation from the indicative application presented here is that the exercise underlines how wide-reaching change in transport supply, as a derived demand, can be. Most of the cases would have positive and/or negative consequences for most of the 17 SDGs. However, it was notable that the two technology substitution cases (HGV-H_2 and EC), as well as avoiding the need for behaviour change – indeed this is the intention of substitution – also offered little in respect of altering the dependencies connected to subsistence. In other words, they make little difference for policy-makers seeking to enhance access to food, water, energy, healthcare, spaces for wellbeing or education. The one exception here was 'employment' and even here the change is through altering the nature, and perhaps increasing the quantity, of employment in the sector, so the relationship is a secondary one, rather than through a spatial change in labour markets due to new transport services.

The three cases offering more 'disruptive' change, deploying more radical technologies and requiring more behavioural change would tend to affect a wider range of dependencies. Here, the changes to the more fundamental SDGs of subsistence could be important: UAVs, for example, might be a way of transporting produce from remote African villages to the nearest market when the roads are impassable in the wet season, whilst MaaS could revolutionise access to healthcare in rural areas.

In all cases, the analyses identify a few clear and strong potential sustainability contributions, such as zero local noxious emissions, but in most cases, the assessments are strongly conditional, subject to better evidence, or down to the net outcome of effects in different directions. For example, reduced GHG emissions are generally subject to the means of generating the energy consumed, with generation occurring in another part of the economic system. Energy efficiency may be enhanced, but it may not be optimised to the extent that would be achieved if external costs are given a higher weighting. Reduced noxious emissions in the local environment may be accompanied by more visual and physical intrusion, and poorer road safety, if reductions in emissions are a factor in perpetuating the dominance of car traffic in cities.

Indeed, these latter uncertainties are examples of the wider principle that what constitutes responsible innovation at any one time is informed by partial and emergent scientific understanding of new technologies and is always socially and political constituted (Stilgoe et al., 2013). The SDG-referenced framework offers the potential particularly to engage stakeholders in addressing, questioning and critiquing that construction, at a strategic level. As such, it can only contribute to a wider architecture for effective governance of innovation, in particular one that embraces scientific study including quantitative evaluations of expected or actual outcomes against targets.

Nonetheless, in order to promote debate, interrogate claims and bring tensions and inconsistencies to the surface, the framework provides business and policy actors in the mobility domain with an accessible operationalisation of the SDGs. In doing so, it facilitates consideration beyond the immediate behavioural or technical implications to consider more fundamentally what major change in the transport sector could mean in terms of maintaining the basics of human life and enabling the human race to actualise a higher level of collective wellbeing.

In practice, the approach would likely function best as an initial scoping analysis tool for policymakers or entrepreneurs wishing to clarify the wider sustainability implications of new business models and technologies, before identifying the most important dependencies in each case for more fundamental analysis. However, they provide a framework to test in broad terms whether sustainable mobility claims are a deep driver or a superficial gloss for mobility innovations.

ACKNOWLEDGEMENTS

The authors acknowledge the projects and funders within which the data used to develop the cases considered in this chapter were originally collected: CAPRI (Innovate UK 103288), CoDeZero and CRAFTeD (UK Engineering & Physical Sciences Research Council), Project PID2020-116040RB-I00 (Spanish Ministry of Science and Innovation), Project GPC-ED431B2022/10 (Xunta de Galicia) and WISE-ACT COST Action CA16222 and its WG3 activities which were co-led by the two first chapter authors.

REFERENCES

Abrar, M. M., Islam, R., & Shanto, M. A. H. (2020, October). An autonomous delivery robot to prevent the spread of coronavirus in product delivery system. In *2020 11th IEEE Annual Ubiquitous Computing, Electronics & Mobile Communication Conference (UEMCON)* (pp. 0461–0466). IEEE.

Ahmed, S., Adnan, M., Janssens, D., & Wets, G. (2020). A personalized mobility based intervention to promote pro-environmental travel behavior. *Sustainable Cities and Society*, *62*, 102397.

Alaerts, L., Van Acker, K., Rousseau, S., De Jaeger, S., Moraga, G., Dewulf, J., De Meester, S., Van Passel, S., Compernolle, T., Bachus, K., Vrancken, K., & Eyckmans, J. (2019). Towards a more direct policy feedback in circular economy monitoring via a societal needs perspective. *Resources, Conservation and Recycling*, *149*, 363–371.

Athanasopoulou, A., de Reuver, M., Nikou, S., & Bouwman, H. (2019). What technology enabled services impact business models in the automotive industry? An exploratory study. *Futures*, *109*, 73–83.

Aurambout, J. P., Gkoumas, K., & Ciuffo, B. (2019). Last mile delivery by drones: An estimation of viable market potential and access to citizens across European cities. *European Transport Research Review*, *11*(1), 1–21.

Bakermans, B. A. (2016, August). *Truck platooning, enablers, barriers, potential and impacts* [Transport, Infrastructure and Logistics Master thesis, Delft University of Technology, The Netherlands].

Berg, I., Rakoff, H., Shaw, J., & Smith, S. (2020). *System dynamics perspective for automated vehicle impact assessment, Volpe National Transportation Systems Center – FHWA Report*. Intelligent Transportation Systems JPO-20-809, US-DOT.

Bezdek, R. H. (2019). The hydrogen economy and jobs of the future. *Renewable Energy and Environmental Sustainability*, *4*, 1.

Black, C., Parkhurst, G., & Shergold, I. (2016). The EVIDENCE project: Origins, review findings, and prospects for enhanced urban transport appraisal and evaluation in the future. *World Transport Policy and Practice*, *22*(1/2), 6–11.

Börjesson, M., Johansson, M., & Kågeson, P. (2021). The economics of electric roads. *Transportation Research: Part C Emerging Technologies*, *125*, 102990.

Bouton, S., Knupfer, S., Mihov, I., & Swartz, S. (2015). *Urban mobility at a tipping point*. McKinsey Center for Business and Environment.

Cabanelas, P., Parkhurst, G., Thomopoulos, N., & Lampón, J. F. (2023). A dynamic capability evaluation of emerging business models for new mobility. *Research in Transportation Business and Management*, *47*, 100964.

Calvert, T., Crawford, F., Parkhurst, G., & Parkin, J. (2022). Perceived accessibility of employment sites by jobseekers and the potential relevance of employer-subsidised Demand Responsive Transport to enhance the commute. *Cities, 130*, 103872.

Calvert, T., Ward, S., Shergold, I., Parkhurst, G., & Jain, J. (2019). Business models being trialled in the shared-ride on-demand niche, and challenges and barriers encountered. In *UTSG 51st Annual Conference*, Leeds, UK.

Caputo, A., Pizzi, S., Pellegrini, M. M., & Dabi⊠, M. (2021). Digitalization and business models: Where are we going? A science map of the field. *Journal of Business Research, 123*, 489–501.

Chechetka, S. A., Yu, Y., Tange, M., & Miyako, E. (2017). Materially engineered artificial pollinators. *Chem, 2*(2), 224–239.

Department for Business, Energy & Industrial Strategy of the UK Government – DBEIS. (2017). *UK greenhouse gas emissions, provisional figures*. Retrieved January 7, 2022, from https://assets.publishing.service.gov.uk/government/uploads/system/uploads/attachment_data/file/695930/2017_Provisional_Emissions_statistics_2.pdf

Eckhardt, J., Nykänen, L., Aapaoja, A., & Niemi, P. (2018). MaaS in rural areas – Case Finland. *Research in Transportation Business & Management, 27*, 75–83.

Energy Monitor. (2021). *Hydrogen tests climate policymakers with its job potential*. Retrieved May 6, 2021 from https://www.energymonitor.ai/tech/hydrogen/hydrogen-tests-climate-policymakers-with-its-job-potential#:~:text=A%20recent%20study%20in%20the,phased%20out%20with%20fossil%20fuels

Epting, S. (2018). Automated vehicles and transportation justice. *Philosophy & Technology, 32*, 389–403.

Ernst & Young Global Ltd. (2017). *Why Sustainable Development Goals should be in your business plan*. https://www.ey.com/en_gl/assurance/why-sustainable-development-goals-should-be-in-your-business-plan

European Commission. (2018). *Measuring progress towards circular economy in the European Union – Key indicators for a monitoring framework*. https://eur-lex.europa.eu/legal-content/EN/TXT/?qid=1516265440535&uri=COM:2018:29:FIN

Eurostat. (2021, February). *Freight transport statistics – Modal split*. Retrieved April 4, 2022, from https://ec.europa.eu/eurostat/statistics-explained/index.php?title=Freight_transport_statistics_-_modal_split

Floater, G., Thomopoulos, N., Docherty, J., Schwinger, P., Mahendra A., & Fang, W. (2017). Accessibility in Cities: Transport and Urban Form. In G. Meyer & S. Shaheen (Eds.), *Disrupting Mobility* (pp 239–273). Lecture Notes in Mobility. Springer, Cham. https://doi.org/10.1007/978-3-319-51602-8_15

Focas, C., & Christidis, P. (2017). Peak Car in Europe? *Transportation Research Procedia, 25*, 531–550.

Genus, A., & Stirling, A. (2018). Collingridge and the dilemma of control: Towards responsible and accountable innovation. *Research Policy, 47*(1), 61–69.

Global Reporting Initiative. (2022). *Most companies align with SDGs – but more to do on assessing progress*. https://globalreporting.org/news/news-center/most-companies-align-with-sdgs-but-more-to-do-on-assessing-progress/

Globisch, J., Plötz, P., Dütschke, E., & Wietschel, M. (2019). Consumer preferences for public charging infrastructure for electric vehicles. *Transport Policy, 81*, 54–63.

Graham, R. (2013). *The future of urban mobility: An overview of current global initiatives and future technologies to help meet the challenges of urban mobility*. A Leasedrive White Paper. Leasedrive, Wokingham.

Greene, D. L., & Parkhurst, G. (2017). *Decarbonizing transport for a sustainable future: Mitigating impacts of the changing climate*. Transportation Research Board, Washington, DC, pp. 30–60.

Greening, P., Piecyk, M., Palmer, A., & Dadhich, P. (2019). *Decarbonising road freight. Future of Mobility: Evidence Review*. Foresight, Government Office for Science. https://assets.publishing.service.gov.uk/government/uploads/system/uploads/attachment_data/file/780895/decarbonising_road_freight.pdf

Haugen, M. J., Paoli, L., Cullen, J., Cebon, D., & Boies, A. M. (2021). A fork in the road: Which energy pathway offers the greatest energy efficiency and CO_2 reduction potential for low-carbon vehicles? *Applied Energy, 283*, 116295.

ICSU, ISSC. (2015). *Review of the sustainable development goals: The science perspective*. International Council for Science (ICSU).

International Energy Agency (IEA). (2022). *Transport sector CO_2 emissions by mode in the Sustainable Development Scenario, 2000–2030*. IEA. https://www.iea.org/data-and-statistics/charts/transport-sector-co2-emissions-by-mode-in-the-sustainable-development-scenario-2000-2030

Jackson, T. (2017). The flying drones that can scan packages night and day. *BBC News*, Retrieved October 27, 2017, from http://www.bbc.co.uk/news/business-41737300

Jaller, M., & Pahwa, A. (2020). Evaluating the environmental impacts of online shopping: A behavioral and transportation approach. *Transportation Research Part D: Transport and Environment, 80*, 102223.

Jittrapirom, P., Marchau, V., van der Heijden, R., & Meurs, H. (2018). Dynamic adaptive policymaking for implementing Mobility-as-a Service (MaaS). *Research in Transportation Business & Management, 27*, 46–55.

JRC. (2019). *The future of road transport - Implications of automated, connected, low-carbon and shared mobility*. Joint Research Center, Ispra: European Commission.

Kim, J., Kim, S., Ju, C., & Son, H. I. (2019). Unmanned aerial vehicles in agriculture – A review of perspective of platform, control, and applications. *IEEE Access, 7*, 105100–105115.

Lawrence, M., Roberts, C., & King, L. (2017). *Managing automation: Employment, inequality and ethics in the digital age*. Institute for Public Policy Research.

Lerner, W. (2011). *The future of urban mobility*. Arthur D Little.

Lewis, A. C. (2021). Optimising air quality co-benefits in a hydrogen economy: A case for hydrogen-specific standards for NO x emissions. *Environmental Science: Atmospheres, 1*(5), 201–207.

Liu, J., Kockelman, K. M., Boesch, P. M., & Ciari, F. (2017). Tracking a system of shared autonomous vehicles across the Austin, Texas network using agent-based simulation. *Transportation, 44*(6), 1261–1278.

Man Truck Germany. (2019). *Platooning in the logistics industry: Researchers see great potential in real operations after tests*. https://www.truck.man.eu/de/en/man-world/man-in-germany/press-and-media/Platooning-in-the-logistics-industry_-Researchers-see-great-potential-in-real-operations-after-tests-372998.html

Marletto, G. (2019). Who will drive the transition to self-driving? A socio-technical analysis of the future impact of automated vehicles. *Technological Forecasting and Social Change, 139*, 221–234.

Milakis, D., Thomopoulos, N., & van Wee, B. (2020). *Policy implications of autonomous vehicles, advances in transport policy and planning*. Academic Press.

Mladenović, M. N., & Haavisto, N. (2021). Interpretative flexibility and conflicts in the emergence of Mobility as a Service: Finnish public sector actor perspectives. *Case Studies on Transport Policy, 9*(2), 851–859.

Nikitas, A., Thomopoulos, N., & Milakis, D. (2021). Environmental impact of Autonomous Vehicles. *Annual Review of Environment and Resources, 46*, 167–192.

Oskam, I., Bossink, B., & de Man, A.-P. (2021). Valuing value in innovation ecosystems: How cross-sector actors overcome tensions in collaborative sustainable business model development. *Business & Society, 60*(5), 1059–1091.

Paddeu, D., & Denby, J. (2021). Decarbonising road freight: Is truck automation and platooning an opportunity? *Clean Technologies and Environmental Policy*, 1–15.

Paddeu, D., & Parkhurst, G. (2020). The potential for automation to transform urban deliveries: Drivers, barriers and policy priorities. In B. van Wee, N. Thomopoulos, & M. Dimitris (Eds.), *Policy implications of Autonomous Vehicles*. Elsevier.

Paddeu, D., Calvert, T., Clark, B., & Parkhurst, G. (2019). *New technology and automation in freight transport and handling systems*. UK Government Office for Science. https://uwerepository.worktribe.com/output/851875

Pagani, M., & Pardo, C. (2017). The impact of digital technology on relationships in a business network. *Industrial Marketing Management, 67*, 185–192.

Pangbourne, K., Stead, D., Mladenović, M., & Milakis, D. (2018). The case of mobility as a service: A critical reflection on challenges for Urban Transport and Mobility Governance. In G. Marsden & L. Reardon (Eds.), *Governance of the smart mobility transition* (pp. 33–49). Emerald Publishing Limited.

Parkhurst, G., (2021). Electric Mobility. In R. Vickerman (ed.), *International Encyclopedia of Transportation* (Vol. 6, pp. 64–72). Elsevier. https://doi.org/10.1016/b978-0-08-102671-7.10615-3

Parkhurst, G., & Clayton, W. (2022). Conclusion: The electric car as a component of future sustainable mobility. In G. Parkhurst, & W. Clayton (Eds.), *Electrifying mobility: Realising a sustainable future for the car* (Chapter 12). Emerald. https://doi.org/10.1108/S2044-9941202215

Parkhurst, G., & Seedhouse, A. (2019). Will the 'smart mobility' revolution matter? In I. Docherty & J. Shaw (Eds.), *Transport matters* (pp. 359–390). Policy Press.

PATH. (2017). *Research: Connected and autonomous vehicles: Truck platooning*. https://path.berkeley.edu/research/connected-and-automated-vehicles/truck-platooning

Portapas, V., Zaidi, Y., Bakunowicz, J., Paddeu, D., Valera-Medina, A., & Didey, A. (2021, July). Targeting global environmental challenges by the means of novel multimodal transport: Concept of operations. In *2021 Fifth World Conference on Smart Trends in Systems Security and Sustainability (WorldS4)* (pp. 101–106). IEEE.

Pütz, F., Murphy, F., Mullins, M., & O'Malley, L. (2019). Connected automated vehicles and insurance: Analysing future market-structure from a business ecosystem perspective. *Technology in Society*, *59*, 101182.

Riggs, W., & Beiker, S. A. (2019). Business models for shared and autonomous mobility. In *Automated vehicles symposium* (pp. 33–48). Springer.

Riggs, W., Beiker, S.A. (2020). Business Models for Shared and Autonomous Mobility. In G. Meyer & S. Beiker (Eds.), *Road Vehicle Automation 7. AVS 2019*. Lecture Notes in Mobility. Springer, Cham. https://doi.org/10.1007/978-3-030-52840-9_4

Rode, P., Floater, G., Thomopoulos, N., Docherty, J. et al. (2017). Accessibility in cities: Transport and urban form. In *Disrupting mobility* (pp. 239–273). Springer.

Sarasini, S., & Linder, M. (2018). Integrating a business model perspective into transition theory: The example of new mobility services. *Environmental Innovation and Societal Transitions*, *27*, 16–31.

Schön, D., & Rein, M. (1994). *Frame Reflection – Towards the resolution of intractable policy controversies*. Basic Books.

Shergold, I. (2019). *Older people and connected autonomous vehicles: An exploration of user needs findings from the flourish project*. Retrieved December 28, 2019, from https://www.researchgate.net/publication/334304159_Older_people_and_connected_autonomous_vehicles_An_exploration_of_user_needs_Findings_from_the_Flourish_Project

Siemens, A. G. (2017). *eHighway – Innovative electric road freight transport*. Information brochure, Munich/Germany.

Singh, K. K., & Frazier, A. E. (2018). A meta-analysis and review of unmanned aircraft system (UAS) imagery for terrestrial applications. *International Journal of Remote Sensing*, *39*(15–16), 5078–5098.

Smith, T., Lewis, C., Faber, J., Wilson, C., & Deyes, K. (2019, July). *Reducing the maritime sector's contribution to climate change and air pollution*. https://assets.publishing.service.gov.uk/government/uploads/system/uploads/attachment_data/file/816015/maritime-emission-reduction-options.pdf

Stilgoe, J., Owen, R., & Macnaghten, P. (2013). Developing a framework for responsible innovation. *Research Policy*, *42*(9), 1568–1580.

Tavasszy, L. A. (2016). *The value case for truck platooning* [Working paper, Delft University of Technology]. https://doi.org/10.13140/RG.2.2.13325.54247

Tilley, S., & Houston, D. (2016). The gender turnaround: Young women now travelling more than young men. *Journal of Transport Geography*, *54*, 349–358.

Turienzo, J., Cabanelas, P., & Lampón, J. F. (2022a). The mobility industry trends through the lens of the social analysis: A multi-level perspective approach. *SAGE Open*.

Turienzo, J., Lampón, J. F., Chico-Tato, R., & Cabanelas, P. (2022b). Electric cars: The Future technological potential. In G. Parkhurst & W. Clayton (Eds.), *Electrifying mobility: Realising a sustainable future for the car (Transport and Sustainability*, Vol. 15, pp. 191–210). Emerald Publishing Limited. https://doi.org/10.1108/S2044-994120220000015012

United Nations (UN). (n.d.). *Making the SDGs a reality*. https://sdgs.un.org/

United Nations (UN). (2014). *The road to dignity by 2030: Ending poverty, transforming all lives and protecting the planet*. Synthesis Report of the Secretary-General on the Post-2015 Agenda.

Van Audenhove, F. J., Korniichuk, O., Dauby, L., & Pourbaix, J. (2014). *The Future of urban mobility 2.0: Imperatives to shape extended mobility ecosystems of tomorrow*. Arthur D Little.

Whittle, C., Whitmarsh, L., Hagger, P., Morgan, P., & Parkhurst, G. (2019). User decision-making in transitions to electrified, autonomous, shared or reduced mobility. *Transportation Research Part D: Transport and Environment*, *71*, 302–319.

Wiegmans, B. W., Hekkert, M., & Langstraat, M. (2007). Can innovations in rail freight transhipment be successful? *Transport Reviews*, *27*(1), 103–122.

World Economic Forum. (2021). *The UN SDGs. How do companies stack up?* https://www.weforum.org/agenda/2021/03/how-aligned-are-un-companies-with-their-sustainable-development-goals/

Zhang, W. (2019). *Planning and evaluation of autonomous vehicles in freight and public transport services* [Doctoral dissertation, KTH Royal Institute of Technology]. http://www.diva-portal.org/smash/get/diva2:1316459/FULLTEXT01.pdf

Zhong, Y., Wang, X., Xu, Y., Wang, S., Jia, T., Hu, X., Zhao, J., Wei, L., Zhang, L. (2018). Mini-UAV-borne hyperspectral remote sensing: From observation and processing to applications. *IEEE Geoscience and Remote Sensing Magazine*, *6*(4), 46–62. https://ieeexplore.ieee.org/document/8573977

CHAPTER 10

CONCLUSIONS: WORKING TOWARDS A SUSTAINABLE ACT

Maria Attard[a], Nikolas Thomopoulos[b] and Yoram Shiftan[c]

[a]*Institute for Climate Change and Sustainable Development, University of Malta, Msida, Malta*
[b]*Department of Tourism & Transport, University of Surrey, UK*
[c]*Faculty of Civil and Environmental Engineering, Technion Israel Institute of Technology, Haifa, Israel*

ABSTRACT

The urgency to act on sustainable transport reached new levels in July 2023 when the temperatures across the globe hit an all-time high, and the promises of Automated and Connected Transport (ACT) to support meeting both the Sustainable Development Goals and reduce the effects of the climate crisis are ever more required. This chapter concludes this volume which aims at the different contributions of ACT to sustainability. It brings together the main themes and perspectives, as well as the challenges highlighted in the previous chapters. The main issues revolve around the effectiveness of environmental sustainability goals for ACT, issues of equity and social sustainability, planning, regulation and enterprise. This work largely represents a collaboration from the Wider Impacts and Scenario Evaluation of Autonomous & Connected Transport (WISE-ACT) Action funded by the COST Programme between 2017 and 2022 which brought together around 200 experts from 42 countries to discuss issues related to Autonomous and Connected Transport implementation.

Keywords: Automated transport; autonomous vehicles; sustainable development goals; research agenda; environmental gains; societal goals

Sustainable Automated and Connected Transport
Transport and Sustainability, Volume 19, 187–194

ISSN: 2044-9941/doi:10.1108/S2044-994120240000019010

WIDER CONTEXT

At the time of writing, in July 2023, the world's highest temperatures were ever recorded. Many disasters related to these extreme temperatures in various parts of the world highlighted, once again, the urgency with which lifestyles and practices need to change in order to reduce the effects of climate change, and focus on building more resilient human environments and systems. More than 2,000 local governments in around 40 countries have declared a climate emergency since 2016, following the 2015 Paris Agreement at COP-21. This includes for example almost three quarters of local authorities in the UK, which have been devising their net zero strategies since. However, there is still a mismatch between local and national governments, since there are local initiatives but no corresponding national ones in certain countries. Following the confirmation of the record high temperatures in the past 120,000 years by the World Meteorological Organization and the EU's Copernicus Earth observation programme, the United Nations Secretary General António Guterres stated: *The era of global warming has ended. The era of global boiling has arrived.* With his words he insisted that this was just the beginning and with immediate climate action, it was still possible to limit global temperature rise to 1.5 °C above pre-industrial levels and avoid the (yet to come) worst of climate change.

The Intergovernmental Panel on Climate Change (IPCC) attribute 15% of the total global emissions to passenger and freight transport in 2019 (IPCC, 2022). A concerted effort in tackling transport emissions would therefore have a significant impact on the global CO_2 burden. Furthermore, regional and local efforts should reflect emission contributions, with the Global North accounting for a larger share of impact and negative externalities for both passenger and freight transport. The opportunities lie in the growing urban populations across the world. The UN estimates over 55% of today's global population is urban and this figure will continue to rise to 68% by 2050 (UN DESA, 2018). The agglomeration effects of urban areas can be channeled to increase efficiency, reduce consumption and implement new technologies that have the potential to assist in the drive towards reducing the burden of emissions. A global, holistic and integrated approach is required to tackle systems, technologies, lifestyles, and the planning and building of our urban environments to reduce the impact on climate change.

There has been considerable investment in technologies over time, some having moderate to high impact on emission reduction whilst others less so. Transport technologies have struggled overall to reduce emissions from the sector as demands for mobility and movement of goods increased unabated. ACT has received considerable attention, and investment over the last few years as a promising technology to tackle transport emissions and meet several relevant sustainability goals. It is unclear whether ACT will reduce or increase the overall Vehicle Kilometres Travelled (VKT), both per capita and per vehicle. It is clear however that as the interest and investment grew, concerns arose out of the need to move beyond the technology fixes as the primary means of addressing sustainability challenges (Creutzig et al., 2018; Thomopoulos & Givoni, 2015).

The WISE-ACT COST Action aimed to investigate the wider impacts of ACT and to describe the best practices on how to evaluate them. It had several objectives[1] including:

- Develop common terminology about ACT across Europe which will be facilitated through an online glossary. It is observed that language and culture across countries and industries play a vital role in understanding key transport notions, so it is useful to develop and use such a glossary to establish a common vocabulary.
- Co-ordinate trials and share know-how across diverse localities to highlight best practices in Europe and beyond. Given that a series of AV trials will take place during the forthcoming years, it is important to coordinate activities and generate comparable outputs based on common research design features.
- Compare simulation results and end user preferences from diverse settings to develop plausible scenarios for ACT deployment. This objective requires deep knowledge of the demand and supply sides of transport networks. Values from diverse trials in simulations are needed to contrast findings and progress from corridor-based to network-based transport and urban planning.
- Roadmap about the wider impacts of ACT deployment to inform policymakers and regulators about the current and future requirements regarding the formulation of relevant standards. Outputs will align with diverse policy targets and may be also used by the automotive industry or local authorities hosting ACT trials.
- Identify and evaluate viable business models for the passenger and freight transport industry. This will include collaboration with local and national authorities to identify value creation activities including those linked with privacy and data management. It will be of value to both the public and private sectors.
- Assess user acceptability and inform the general public and key stakeholders about the opportunities and risks of ACT. This will facilitate formal and informal debates to take place and can increase the acceptability of ACT.

These objectives were met through (i) the building of a community fostering scientific dialogue, knowledge exchange, and the development of a consolidated thematic framework about the wider impacts of Autonomous and Connected Transport, (ii) bridging distinct scientific fields such as legal informatics, transport economics and engineering, and (iii) acting as a transnational stakeholder platform providing evidence-based recommendations about ACT deployment. This volume is one of the outputs of this extensive network of collaborators, over 200 experts from 42 countries in Europe and beyond.

Moreover, to meet objectives WISE-ACT launched a large survey with more than 4,000 respondents in more than 20 countries, which was among the first to address the user perspective at such an international scale (Andrei et al., 2022; Etzioni et al., 2020; Hamadneh & Esztergár-Kiss, 2021; Kyriakidis et al., 2020,

[1]https://wise-act.eu/about/objectives/

Polydoropoulou et al., 2021). A key contribution of the WISE-ACT survey was that it also included respondents outside the European Union, e.g. Iceland, Israel, and Montenegro. That WISE-ACT survey formed the basis for a large part of the Special Eurobarometer survey (European Commision – Joint Research Centre, 2020), which was conducted on behalf of the Joint Research Centre of the European Commission and included 27,656 respondents across 28 European countries.

Similarly, to the WISE-ACT survey, findings of this Special Eurobarometer survey showed that 60% of respondents across the European Union (EU) have heard, read or seen something about Automated Vehicles in the previous 12 months. However, there was major variation between countries, which further increased within each country. The highest awareness levels were reported in The Netherlands (93%), Sweden (91%), Denmark (85%), and Estonia (83%), whereas the lowest levels were reported in Italy (36%), Poland (34%), Romania (29%), and Bulgaria (22%). Despite this being only an ACT awareness snapshot of the general public, it clearly demonstrates the wider underlying challenges of informing users about the potential ACT benefits in addressing climate change and meeting SDGs. On the one hand, it is optimistic that awareness levels have been constantly increasing since 2020, so the general public is becoming more aware of the potential advantages of ACT to meet broader environmental, social and economic goals. On the other hand, it seems that some experts have become less optimistic about ACT benefits (Kroesen et al., 2023). The economic case of ACT has been fiercely debated, but is still very uncertain depending on the level of ACT adoption. Therefore, the following sections focus on the environmental and social pillars of sustainable development.

CONTRIBUTING TO ENVIRONMENTAL GAINS

Three chapters in this volume contributed directly to the discussion related to the environmental gains of ACT. There are still many debates concerning these issues. The research so far has identified three main dependencies upon which ACT would result in a positive impact on the environment. The chapters in this volume attempt to quantify, assess and review sustainability, but similar to other researchers, still contend that a successful outcome of ACT deployment will depend on the levels of sharing of the fleets and infrastructures that are developed. The unsustainable energy usage linked to individual travel and the related emissions are highlighted, and as such, it is only through shared-based approaches and business models that sustainable and long-term gains will be achieved. The level of automation will also have an impact. The chapter by Dadashzadeh, Agriesti, Sadid, Elvarsson, Roncoli, and Antoniou evaluates sustainability through simulations of Autonomous and Connected Transport. They develop a list of Key Performance Indicators for the assessment of policy decisions with respect to traffic efficiency, safety, energy consumption and emissions, travel behaviour and land use.

It is evident that there is still room for further research in the area with the mobility transition being defined by Nikitas as potentially lengthy, easy-to-go-wrong and complicated. His chapter in this volume raised a number of questions and concerns about the likely ACT impacts and side effects. These concerns require strategic and interdisciplinary approaches that combine the technology development with necessary shifts in policy and travel behaviour, and which should go beyond currently considered options (MaaS Alliance, 2023), for example, sustainable multimodal transport through Mobility-as-a-Service (MaaS). Indeed, there is already evidence, some of which is suggested in this volume, that ACT will not resolve all traffic problems which generate environmental, but also economic and social problems.

Lastly, the third and probably most difficult issue to predict is the level of new vehicle kilometres which ACT will attract from active travel modes, public transport and new mobility demands and aspirations, facilitated for those currently not willing or unable to drive. Only if done in large proportions and in shared mode, there could be significant gains. It is also clear that much of the implications studied so far do not go beyond the Global North and predominantly urban centres, with very little known with respect to the rural or Global South situations. Given the constantly deteriorating air quality conditions in urban areas, it is crucial to conduct more accurate studies about the ACT impact on pollution in both urban and rural areas, since rural areas are often affected more severely by climate change.

THE SOCIAL IMPLICATIONS

The WISE-ACT community was primarily inspired by the potential of the so-called 'social implications' of ACT. These concerns stem from overtly simplistic approaches and analysis that justify automation. Such claims, as the ones made in the 2017 RAND report, suggest that AVs should be deployed '*when their safety performance is only moderately better than that of the average human driver*' (Kalra & Groves, 2017, p. 10). The community of researchers acknowledged the opportunities as presented in such 'grey' literature but also questioned their contribution to policy. Despite advances in academic research in the field of automation in transport, social or 'wider' implications remain a relevant and much needed area of research.

The ACT-associated opportunities and risks have been the subject of the WISE-ACT Action, with a review of literature compiled as part of the Action Thematic reports. The social implications beyond road safety are many, with a good number of recent academic works focusing on user-centric approaches to study effects on equity and inclusion. The chapter by Beyazit, Soh and Martens brought together this literature and drew up policy recommendations for governance.

ACT concerns are also reflected upon by Schallauer, Soteropoulos, Cornet, Klar, and Fürdös in their chapter dealing with the regulatory frameworks at national levels across Europe and the key aspects for a sustainable implementation. This chapter is critical to the broader discussion as it makes it clear that such frameworks at the national level are still lacking. Is this a case of technology marching ahead of, and faster than, regulatory or administrative processes?

This concern arose many times in previous transport (technological) development phases such as the rise of individual/shared mobility applications, and more recently with the deployment of micromobility modes across cities, without the necessary regulatory frameworks to control and safely manage these more (cyber) libertarian approaches to systems and services (see, for example, Aarhaug et al., 2023; Attard et al., 2016; Moran et al., 2020). Paris has been a remarkable example, where e-scooters were first introduced in 2018 and then were first banned in 2023. The questions regarding employment, security, safety, and violation of contracts and issues with equity have been debated among scholars (Epstein, 2015). There is therefore increased pressure on governments to balance out the innovation that seeks to 'solve' problems with the current transport systems and societal concerns with growth and sustainable economic development.

With new developments in AI, as recorded at the time of writing in 2023, there are greater calls for debate and action to embed ethical and security challenges in regulatory frameworks. These, as suggested by authors like Milakis et al. (2020), should work towards more responsible innovation within regulatory frameworks that protect participants and the general public.

THE FUTURE AGENDA

The WISE-ACT community remains active in the broader ACT field. The challenges that arise out of the new technologies are still the subject of ongoing research undertaken at local, national, and international levels. This volume has tackled a number of these challenges and set them aside from the objectives of the UN Sustainable Development Goals, in an attempt to evaluate the impacts (both positive and negative) of such a transition.

Further work however is required to inform the debate and provide independent and evidence-based research to support practitioners and policy makers. The idea of exploring the wider implications of ACT is still very much relevant, even after four years of discussion, networking and publishing in the field. The future agenda for ACT deployment must include a broader understanding of the implications on the environment, society and the economy. Adopting a quadruple or even quintuple helix framework as suggested in Chapter 1 would certainly be beneficial and would provide a valuable foundation for such a multifaceted challenge as the evaluation of ACT deployment. All this necessitates a strong(er) regulatory framework within which constraints and control measures are set to increase security, safety and protection of the most vulnerable within society. These might be high-level aspirations from this volume, however, it is evidently clear that a risk-free and unmitigated deployment of automation technology in transport is no solution to the current climate crisis, as well as to the negative social and economic impacts of contemporary transport systems.

A clear recommendation which has been constantly raised by experts in different fields and in different countries is the need for ACT training and education. This was the theme of the 4th WISE-ACT 2019 workshop held in Riga and it reaffirmed the broad consensus about it. As shown by the WISE-ACT and Eurobarometer survey findings, awareness levels vary significantly across countries but have been increasing since 2020. Yet, educating the general public has been an area which has received less attention from practitioners and policymakers. ACT opportunities and risks linked with travel options for older, younger populations, disabled or other vulnerable populations require further attention. Equally, future research should focus not only on the pressing technological ACT challenges such as wireless network latency, but also on wider transport network issues such as energy use. Along the same lines, the ACT uncertainties surrounding data management or socio-economic issues such as privacy and other human rights affected by AI and digital technologies, should become the focus of future intradisciplinary research.

Sustainable ACT deployment that factors in basic human needs and recognises the earthly boundaries with respect to pollution and negative impacts emanating from unsustainable resource use are just two of the more pressing concerns. Sustainable economic growth that is supported by truly sustainable means of transport alongside the protection of values such as equity and access are needed. This volume has contributed to this discussion, and hopefully has broadened the research agenda for future work in this field. There is a strong belief in the promise of ACT. This potential however is only going to materialise if stakeholders embrace sustainability goals and develop deployment frameworks that resonate with the need for reducing impact, building more resilient cities and ensuring equity in governance and regulation.

REFERENCES

Aarhaug, J., Fearnley, M., Liland Hartveit, K. J., & Johnsson, E. (2023). Price and competition in emerging shared e-scooter markets. *Research in Transportation Economics*, *98*, 101273. https://doi.org/10.1016/j.retrec.2023.101273

Andrei, L., Luca, O., & Gaman, F. (2022). Insights from user preferences on Automated Vehicles: Influence on socio-demographic factors on Value of Time in Romanian Case. *Sustainability*, *14*, 10828. https://doi.org/10.3390/su141710828

Attard, M., Haklay, M., & Capineri, C. (2016). The potential of Volunteered Geographic Information (VGI) in future transport systems. *Urban Planning*, *1*(4), 6–19. https://doi.org/10.17645/up.v1i4.612

Creutzig, F., Roy, J., Lamb, Azevedo, I. M. L., Buine de Bruin, W., Dalkmann, H., Edelenbosch, O. Y., Geels, F. W., Grubler, A., Hepburn, C., Hertwich, E. G., Khosla, R., Mattauch, L. Minx, J. C., Ramakrishnan, A., Rao, N. D., Steinberger, J., Tavoni, M., Ürge-Vorsatz, D., & Weber, E. U. (2018). Towards demand-side solutions for mitigating climate change. *Nature Climate Change*, *8*, 260–263. https://doi.org/10.1038/s41558-018-0121-1

EC-JRC (European Commission – Joint Research Centre). (2020). *Expectations and Concerns of Connected and Automated Driving, Special Eurobarometer 496 Report, Wave EB92.1, Joint Research Centre*. European Commission. https://europa.eu/eurobarometer/surveys/detail/2231

Epstein, R. (2015). *The libertarian: "Uber and innovation"*. Hoover Institution. http://www.hoover.org/research/libertarian-uber-and-innovation

Etzioni, S., Hamadneh, J., Elvarsson, A., Esztergár-Kiss, D., Djukanovic, M., Neophytou, S., Sodnik, J., Polydoropoulou, A., Tsouros, I., Pronello, C., Thomopoulos, N., & Shiftan, Y. (2020). Modelling cross-national differences in Automated Vehicle acceptance. *Sustainability*, *12*, 9765. https://doi.org/10.3390/su12229765

Hamadneh, J., & Esztergár-Kiss, D. (2021). The influence of introducing autonomous vehicles on conventional transport modes and travel time. *Energies*, *14*(14), 4163. https://doi.org/10.3390/en14144163

IPCC. (2022). *Working Group III contribution to the Sixth Assessment Report: Summary for policy makers*. Intergovernmental Panel on Climate Change. https://www.ipcc.ch/assessment-report/ar6/

Kalra, N., & Groves, D. G. (2017). *The enemy of good. Estimating the cost of waiting for nearly perfect automated vehicles*. RAND Corporation. https://www.rand.org/pubs/research_reports/RR2150.html

Kroesen, M., Milakis, D., & van Wee, B. (2023). Automated vehicles: Changes in expert opinions over time. *Transport Policy*, *136*, 1–10.

Kyriakidis, M., Sodnik, J., Stojmenova, K., Elvarsson, A., Pronello, C., & Thomopoulos, N. (2020). The role of human operators in safety perception of AV development – Insights from a large European survey. *Sustainability*, *12*, 9166. https://doi.org/10.3390/su12219166

MaaS Alliance. (2023). *MaaS Driving Sustainability, Position Paper in collaboration with 5T-Torino and Connected Places Catapult, Brussels*.

Milakis, D., Thomopoulos, N., & Van Wee, B. (2020). *Policy implications of autonomous vehicles, Advances in Transport Policy and Planning Series*. Academic Press – Elsevier.

Moran, M. E., Laa, B., & Emberger, G. (2020). Six scooter operators, six maps: Spatial coverage and regulation of micromobility in Vienna, Austria. *Case Studies on Transport Policy*, *8*(2), 658–671. https://doi.org/10.1016/j.cstp.2020.03.001

Polydoropoulou, A., Tsouros, I., Thomopoulos, N., Pronello, C., Elvarsson, A., Sigursson, H., Dadashzadeh, N., Stojmenova, K., Sodnik, J., Neophytou, S., Esztergár-Kiss, D., Hamadneh, J., Parkhurst, G., Etzioni, S., Shiftan, S., & Di Ciommo, F. (2021). Who is willing to share their AV? Insights about gender differences among seven countries. *Sustainability*, *13*, 4769. https://doi.org/10.3390/su13094769

Thomopoulos, N., & Givoni, M. (2015). The autonomous car: A blessing or a curse for the future of low carbon mobility? An exploration of likely vs. desirable outcomes. *European Journal of Futures Research*, *3*(1), 1–14. https://doi.org/10.1007/s40309-015-0071-z

UN DESA. (2018). *2018 Revision of the world urbanization prospects*. Population Division, United Nations Department for Economic and Social Affairs. https://population.un.org/wup/publications/Files/WUP2018-Report.pdf

INDEX